ADVANCE PRAISE FOR

"Gerry is one of those rare individua... business acumen are only eclipsed by the ... his heart and the depth of his compassion for those who need a hand up. A brilliant leader and one of the most humble and generous human beings I've ever had the pleasure of knowing. His recipes for business and life are laid out clearly in *I'm Just Gerry*."

Mark J. Chipman, Executive Chairman and Governor
Winnipeg Jets Hockey Club

"*I'm Just Gerry* tells an amazing story of doing business the right way … the Price Way. Gerry's story is a gift to any leader who wants to find and follow a principled path to success while leaving a positive legacy. For those that want to build a lasting enterprise and are prepared to think long term and seriously out-serve and outlast the competition, the recipe and lessons are clear and there to follow in *I'm Just Gerry*."

Dayna Spiring, President and CEO
Economic Development Winnipeg

"*I'm Just Gerry* is as honest, sincere, and transparent a book as they come. Just like Gerry is. Price is an amazing business with an incredible history having evolved over the years with purpose and vision, commitment to employees and customers, and plain old-fashioned hard work. There is a lot to be learned here from a very special leader and his company."

Paul Soubry, President and CEO
NFI Group Inc. (New Flyer Industries)

"Very few people are awarded both an Honorary Degree and a Distinguished Alumni Award from the University of Manitoba, but Gerry Price has been because he is, in a word, exceptional. He is a gifted engineer and brilliant entrepreneur whose primary concern always seems to be the wellness of others. The University of Manitoba can easily point to our engineering faculty that bears his name to show his impact—he does so much to build communities and create a better future. If you're looking for someone to emulate or learn from, I'd say just start with Gerry. I am thrilled to see Gerry's story and the Price Way captured and preserved through this book—the stories and lessons shared will serve as a valuable resource and inspiration for many generations to come."

Michael Benarroch, President and Vice-Chancellor
University of Manitoba

"Gerry's vision for Price has always been to be number one. He has invested in that vision through good times and doubled down during bad times. He advocates that you must build an adequate foundation to grow. I took Gerry's philosophy to heart when we bought the Tom Barrow business in 2015. And it worked! I wish all of our supplier partners had this vision and commitment. I encourage you to read this book and learn from one of the best."

Mike Shea, President and CEO
Tom Barrow Company

"People who make a real difference in our world can take many forms—those who quietly contribute, those who role up their sleeves and help, and those who take the lead to forge the path forward. Gerry is one of those exceptional individuals who has made a difference in our world in all three ways. He and his family selflessly give back, he leans in when called on to help, and in business and community he leads with vision, passion, and conviction."

Paul Mahon, President and CEO
Canada Life and Great-West Lifeco

"I think it's fair to say that Gerry has helped shape Winnipeg as much as Winnipeg shaped who he is as a business leader: resilient, resolute, and willing to take a risk in order to move forward. Gerry and I have known each other for 40 years, going back to the mid-1980s when our family held an ownership interest in E.H. Price. What impressed me about Gerry from the start was his determination to lead his family's manufacturing company into the future and his unwavering stance on keeping its headquarters here in Winnipeg while growing its presence internationally. That commitment is not only a testament to Gerry's passion for Winnipeg; it is a reflection of the hardy, homegrown values the city instilled in him."

Hartley Richardson, Executive Chair, President, and CEO
James Richardson & Sons, Limited

I'M JUST
GERRY

A Wood Dragon Book

I'M JUST
GERRY

*like an ant on a
blueberry pie*

BUILDING A FOREVER COMPANY
THE PRICE WAY

Written by Rob Wozny
Afterword by Gerry Price

A Wood Dragon Book

I'm Just Gerry
Like an Ant on a Blueberry Pie
Building a Forever Company the Price Way

Copyright © 2023 Price Industries Limited

Written by Rob Wozny
Foreword by Harry Ethans
Afterword by Gerry Price

Cover design: Rita Yermolenko
Interior Book Design: Christine Lee

ISBN: 978-1-990863-33-2 (Hardcover)
ISBN: 978-1-990863-32-5 (eBook)
ISBN: 978-1-990863-31-8 (Paperback)

Contact information:
Author: Rob Wozny www.robwozny.com
Price Industries: info@imjustgerry.com

Printed in Canada

Net proceeds of this book go to the
Price Family Foundation

DEDICATION

This book is dedicated to our past, present, and future
employees, whose extreme service to customers and each
other, operating in the Price Way, has created the growth and
success we've witnessed to date and hope to see continue into
the future. It is also dedicated to our many long-term and more
recent customers, whose success in their local markets is the
only reason we've been successful.
Finally, and most importantly, this book is dedicated
to my wife, Barb, the love of my life, my best friend
and soulmate, who has been my life partner since
high school.
Gerry Price

For entrepreneurs, innovators, and leaders
seeking affirmation that there are no shortcuts
on the path to finding success in business.
Rob Wozny

CONTENTS

Foreword
Harry Ethans

At Price Industries' headquarters in Winnipeg, Manitoba, Canada, Gerry Price is surrounded by dozens of invitees, including members of the business elite, influential community supporters, and his team of business leaders, some of whom have traveled from other provinces and from the US. The Bankers' Orientation is Gerry's annual "state of the union" presentation, which showcases how Price Industries will achieve and maintain its ambitious growth trajectory. The event was launched in 1987 as an in-person checkpoint to keep financial institutions, key suppliers, and customers informed as Price relied heavily on their support during those heady US-expansion days. Just like the thousands of others who have seen Gerry speak passionately at this event over the decades, my first Bankers' Orientation in spring 2002 anchored my admirable first impression of him. Over the past 20 years, that impression of Gerry has never wavered as he has consistently lived his life the Price Way. I have enjoyed a personal and professional relationship with him, earning his trust as an inner-circle business advisor and becoming a longtime friend.

As I was getting to know Gerry, I was coming out of a long-term business relationship with Israel "Izzy" Asper—a made-in-Manitoba media mogul—where I was fortunate enough to have a front-row seat and an opportunity to participate in some of the most exciting and game-changing transactions in the broadcasting and print industries. Izzy had a dream to build the third national television network in Canada and to create a far-flung global media empire. No one thought it could be done from a prairie outpost, but just like Gerry, Izzy and the CanWest team proved the naysayers wrong.

At his Bankers' Orientations, Gerry has always shared in intricate detail his growth plans for Price Industries, so it is no surprise to me that you are reading *I'm Just Gerry* today. It is the natural next step in the evolution of Gerry's journey—one that serves to educate, inform, and inspire employees and external stakeholders in Price Industries' pursuit of building a legacy organization. So committed is Gerry to sharing his journey for the benefit of others that he has drafted a comprehensive afterword for *I'm Just Gerry*. As only Gerry can deliver, the afterword is a meticulous timeline narrative that complements the preceding manuscript. Both components of the book will tantalize the reader's imagination and satisfy the fact finder's curiosity to know exactly how Gerry did what by when.

More personally, *I'm Just Gerry* aligns instinctively with Gerry, an educator at heart. He is genuinely keen about people learning—specifically learning by doing—and *I'm Just Gerry* is demonstrable proof of how the Price Way can be an invaluable resource, vividly detailing the highs and the lows of the history of the growth of Price Industries, most notably the US-expansion years in the early 1990s. I commend him for sharing his story, both at my first Bankers' Orientation and now with *I'm Just Gerry*.

I relate intimately to Gerry's learn-by-doing approach, and I have every confidence that readers will too. It was how I was taught corporate development and finance techniques during my formative years with Izzy and his business partner, Gerald W. Schwartz. I worked with Izzy for close to 20 years in various senior capacities as CanWest Global grew to become an international media conglomerate with outlets around the world. I never thought I would become capable of leading major corporate transactions, acquisitions, and mergers. I learned by watching, participating and, at times, making mistakes. The critical components of my development were taking responsibility and learning from any missteps. Consistent with Gerry's leadership style, he does not harshly criticize his team when errors or mistakes inevitably take place. In fact, he is the first person to congratulate someone for taking ownership of a presenting problem and recommending a solution.

These solutions often benefit the customer at Price's expense, which at first blush seems counterintuitive. However, this unselfish and right-minded approach pays enormous dividends in the long run. As documented throughout *I'm Just Gerry*, Gerry would rather have leaders understand what went wrong, take steps to ensure the misstep does not happen again, learn how to become better from the experience, and then get back behind the wheel.

For Gerry, growing a business is an iterative process. If you're designing new products, or you're coming up with a different theme or concept, don't wait until you've got the perfect product—get a prototype, test it in the lab, and make sure it is working properly. Tweak it if you have to, produce it, and get it out. There's going to be a version two and a version three, and a version seven. But if you sit on it, it is never going to get out to market. Most companies do not possess the resilience and grit to wait out the costly and time-consuming process that is necessary to innovate and break into new markets. More importantly, Price Industries pays close attention to the marketplace acceptance of new-product introductions and heeds the comments and advice provided by its most trusted sales representatives to refine the devices as required.

What also sets Gerry apart from most accomplished business leaders is his commitment to transparency in communicating the immense value of service. At that impressionable first Bankers' Orientation, I was incredulous as I watched him stand up in front of his current funders, prospective bankers, accountants, and members of the business community to not only tell his story with a historical perspective, but wax poetic on bold prognostications. I was taken aback by Gerry's open and frank approach. My corporate training was to be far less candid and much more circumspect when providing forward-looking predictions of a financial nature, especially in a quasi-public setting. But there he was, unwavering in his vision, expounding how Price Industries was going to achieve such lofty goals, including offering up revenue forecasts and competitive positioning strategies. I thought it was refreshing, albeit a little bit risky; it could backfire if his projections were not substantially realized. In that moment, I was struck by how unique a leader Gerry was; confident and self-assured, but certainly not brash or arrogant. He did not oversell, but simply extrapolated how he saw things and the path it would take to get there.

Most companies, especially publicly traded companies, are driven by quarterly earnings. Those companies may have a mission and vision statement, professed values and principles, and even tenets of their own—all that speak to wanting to produce great products and services and do the right thing for their customers. But at the end of the day, many of those companies will be forced to focus on short-term results in order to satisfy market expectations and those of their stakeholders.

That is not the case with Gerry.

His life's work is to be of service to others, not just to his customers, but to his employees and to the people in the communities where Price Industries does business. It's going above and beyond, treating people the way he would want to be treated—the Golden Rule. It has always been purpose over profits for Gerry—being of service above all else.

Introduction
Why "I'm Just Gerry"

Determining a book's title is often one of the more challenging tasks in a manuscript's development. Finding that near-perfect phrase that succinctly, accurately, yet creatively sums up what the book is all about can be a contemplative and an unnerving process, complete with endless suggestions and iterations of those suggestions—and at the end of it all, many authors still do not find that desirable catchall title. That was not the case with *I'm Just Gerry: Like an Ant on a Blueberry Pie.*

For Gerry Price, who will be referred to as just Gerry from this point forward in the book, is a natural storyteller, armed with an almost endless supply of well-timed sayings, or "Gerryisms." The one simile he frequently uses to sum up his personal ambition and drive for something much bigger than himself comes down to one folksy yet simple phrase: "Like an ant on a blueberry pie." Gerry metaphorically sees his journey like that of an indefatigable ant, searching for sustenance, that instinctively finds its way to a blueberry pie, musters the disproportionate amount of strength that ants are known for, and pulls one blueberry out of the hundreds baked into the pie. All Gerry wanted was just one blueberry, whetting his appetite symbolically for a full slice of the market (air distribution), growing Price Industries from a prairie powerhouse to an international presence in the heating, ventilation, and air-conditioning (HVAC) industry.

During the dozens of interviews conducted for this book, most of the interviewees had some congenial exposure to Gerry's affable blueberry pie metaphor, confirming that this title was indeed the intuitive and correct choice. After three decades leading Price Industries and its affiliate companies, Gerry is as committed to getting his next slice of blueberry pie as he was to the first—building foundation and legacy for generations to come.

In simpler terms, building foundation and legacy means investing in the right people, products, and service, supported by continuous investment of time, money, and resources—most notably when times get tough, including unpredictable circumstances such as a recession or a pandemic. According to the Price Way, if the return on investment takes considerably more time and money, so be it, as the knowledge gained and the lessons learned will serve greater, more profitable endeavors in the years ahead.

The need for *I'm Just Gerry*

In 2024, Price Industries commemorates 75 years in business. That milestone is no insignificant feat, especially following a pandemic that wiped out numerous venerable companies across a variety of industries worldwide. *I'm Just Gerry* captures Gerry's remarkable 37-year journey from becoming President of Price Industries in 1986 to creating and expanding the Tenets that comprise the Price Way, taking the unfathomable risk that put the company's and his family's assets in jeopardy, and leading the company to unprecedented growth.

While Gerry persevered through all kinds of risk and never relented on his growth ambitions, what matters most to Gerry now is ensuring that the Price Way of doing business remains foundational for the company's leaders of today, tomorrow, and beyond. For anyone who has a stake in serving Gerry in his quest, this book is for you.

In the spirit of service, this book is also for anyone seeking another way—the Price Way—of doing business. It is essential to note the intention of *I'm Just Gerry* is not to inculcate the 13 Tenets as the only way, but to introduce readers to another route to consider as they make their way through the complexities of the business world. Accordingly, *I'm Just Gerry* articulates how one Canadian company, anchored by 13 guiding Tenets, succeeded despite extraordinary odds. Each Tenet aptly serves as a chapter, unveiling all the key ingredients that are baked into Gerry's blueberry pie, so to speak, and humbly serves the greater good to anyone looking for entrepreneurial inspiration. *I'm Just Gerry* is a business story of hard lessons learned, setbacks, struggles, dogged determination, the pursuit of innovation, and the privilege to truly serve.

Backed by the learn-by-doing Price Way best practices and reinforced by decades of trials and tribulations, the Tenets showcased in *I'm Just Gerry* go well beyond mere principles. Covered in greater detail throughout the book and shared here as a precursor to the book-reading journey ahead, Gerry's business essentials that are covered in the Tenet chapters include the following:

- Build a company from greenfield start-ups
- Know your critical factors for success
- Recognize incremental margin must be greater than incremental overhead
- Grow the company discreetly and with integrity
- Take on small and medium jobs, balanced by larger projects
- Remain viable through periods of growth without creating detriment to the customer
- Embrace people at every level of growth and invest continually in their development

For anyone who seeks inspiration from a company that has defied the crazy odds of not only succeeding but staying relevant more than seven decades after its founding, this book is an extension of Gerry's commitment to service. His summation of what he hopes readers gain from the book includes: "I'd like to see us pass along in this book the inspiration for people to take this process, and this ridiculous path I took, to gain their own knowledge. And I say 'ridiculous' because I never dreamed this path would lead to a business of this success."

Take from *I'm Just Gerry* what you will, and we wish you nothing but success in the process.

The strength of an anvil

Over the decades, Gerry's company has evolved into a series of supporting businesses that facilitate consistent growth and sustainability. Officially known as Anvil Management, this corporate entity owns Price Industries (and divisions such as Noise Control, Antec Controls, Price Electronics, SolutionAir, and E.H. Price) and a series of affiliate companies (including APEL Extrusions Ltd. and AROW Global).

Anvil was formed in 1985, shortly before it became necessary for Gerry to buy out exiting shareholders who were not fully on board at the onset of his ambitious US expansion (the slice of blueberry pie). The name was chosen by Gerry and his wife, Barb, as a symbol of strength.

For clarity and ease of readership, Anvil Management will be referred to as **Anvil** throughout *I'm Just Gerry*. When referencing Price Industries, the core HVAC company, which makes up over 70% of Anvil, we will truncate the name to **Price**.

"I'm just Gerry"

At the April 2022 edition of the annual Bankers' Orientation, the vibe in the room had a rock-show feel to it, humming with anticipation. Gerry's presentation was 45 minutes long, and after 30 minutes of heaping effusive praise on those who had helped to grow the company to its historic levels, he slowly turned away from the large projector screen behind him and paused. With humility, he looked intently at the crowd and self-deprecatingly offered his assessment of his own contributions to his company's growth: "And me, well, *I'm just Gerry*."

The consensus from the stakeholders featured in *I'm Just Gerry* easily counters that Gerry is a lot more than *just* Gerry: his legendary stamina and unwavering commitment to exceptional service, relentless innovation, and continual growth have almost single-handedly won over the hearts and minds of contractors, the engineering community, and top US sales representatives, along with other business leaders, bank executives, and even fierce competitors, allowing Price to break through into the competitive US market.

As referenced often throughout *I'm Just Gerry*, a sales representative, known more conversationally within Price's ecosystem as a "rep," is not just one person but a team of sales professionals working for a sales representative company. A rep company buys and sells products made by manufacturers like Price, working with contractors and engineers in their markets who design and build commercial buildings. Buying products for their customers from Price, the reps, in turn, sell them to the contractors at a markup.

With total annual sales in 2022 of more than $1 billion spread across 11 countries and an employee force of more than 4,500, many of Anvil's companies have also become industry leaders in their respective fields. Furthermore, the company's commitment to and investment in innovation over the years have led to steady growth since 1985.

From the kitchen table to the boardroom table

Many of the world's most successful brands and businesses come from some sort of proverbial humble beginnings, usually with a vision and mission statement handwritten at a kitchen or dining room table on whatever was readily available to capture the words into statements that would stand the test of time. "I took on the challenge of creating the company profile and marketing materials. In fact, our vision and values evolved into the Tenets we have today. We had an early version of them in the fall of 1986, starting with eight principles that evolved over three decades into the 13 Price Way Tenets that guide us today," says Gerry, recalling that seminal moment in the company's history. Those formative Tenets were drafted around Gerry's dining room table in the late 1980s with someone whom Gerry draws strength and character from—his wife, Barb.

While the decades have passed, Barb's memory of Gerry drafting the early version of the Tenets remains as vivid as though it happened yesterday. "Absolutely. I can see our dining room table as if it were right here, and when I read his vision statement for the first time, I said, 'That sounds pretty lofty.'"

Looking back before looking forward

To get a better understanding of the Price Way's Tenets and the man who authored them, it is important to understand the formative years of the company.

Gerry's father, Ernie, started E.H. Price Ltd. in 1949 as a one-person company, selling HVAC products made by others. Gerry joined E.H. Price Ltd. (now Price Industries Ltd.) in 1977, ascending to the role of company president in 1986. While they shared the Price surname, Gerry and his father never worked together in a formal leadership structure; Ernie retired five years before Gerry joined the company. Nonetheless, Ernie was pleasantly surprised when Gerry came on board, given Gerry's initial career path in research. Recruiting Gerry into Price was always a hopeful outcome of the company's succession plan, and assigning him various leadership roles is accredited to then-President Gerry Law, whom Gerry would eventually succeed. "But what Gerry Law gave me was the gift of independence. In return, I always kept him informed on what I was doing," says Gerry.

Officially, Gerry became President of Price at the age of 38, though his exposure to the company's operations began in earnest in his teenage years. Sharing the company's namesake was where the preferential treatment ended, instilling in Gerry a sense of service and a desire to get every task done to the best of his abilities, no matter what that task involved. "First of all, as a kid, I worked in the plant in Winnipeg. I did all kinds of odd jobs around the plant. I painted the walls and did maintenance work. Before I was 16, I was cleaning fish tanks in my dad's office and unloading scrap aluminum from the extrusion press and putting it into a truck to take the scrap away for remelt."

While these were never the most glamorous tasks, Gerry claims he developed an appreciation and understanding that every chore has a purpose and had to be done. "I cut the grass of not just our own family, but my grandparents, my aunts, and uncles. I just did it if it needed to be done. I learned how to dig a hole efficiently with a spade thanks to my World War II spade from Dad. So I know the importance of the ergonomics of the spade because I dug a lot of different holes in very hard clay."

A learn-by-doing approach

Possessing an innate understanding of how things work

came to Gerry. This understanding was developed further by his formal education in engineering, which included bachelor's and master's degrees from the University of Manitoba and a PhD from Lehigh University in Pennsylvania.

Raised by Depression-era parents who had children later in life, Gerry expresses how he was literally left to his own devices with practically no parental supervision for the bulk of his childhood and teenage years. With his parents preoccupied with work responsibilities that dominated the discretionary time of most families of that era, Gerry shares how that void of parental guidance left him with countless hours to fill his time, leading to his early engineering interests. "I learned how to use my hands to do jobs. I built little rockets out of tubes. We made our own gunpowder with saltpeter and sulfur. I built various things electronically, like radios. I didn't understand electronics, but I could follow the colored wires, and I tried to make things work."

As a teenager, Gerry put his carpentry skills to work building a small hydroplane boat in his basement, which would benefit his developing romantic interest with a girl he was smitten with at high school. With the boat tested and complete, Gerry piqued the interest of his high school crush to go for a cruise in his new boat on the Red River. He launched from East Kildonan in Winnipeg, close to where he lived, and traveled to North Kildonan to pick up Barb from her family home. They traveled to Lockport, 13 miles north.

Now, more than 50 years later, Barb fondly remembers that adventurous ride as the launch point going from high school sweethearts to lifelong partners: "To be picked up in a little hydroplane behind your house for a picnic? That's pretty darn romantic, right? Until I got in it and realized that underneath the flotation device Gerry gave me to sit on, the boat shell was just three-quarter-inch plywood. To sit together on that, sharing it as we were hit by waves—you know how uncomfortable that was?" She laughs, "They're just wonderful memories."

"Ace in the hole"

A guidance counselor in high school took notice of Gerry's interest in mathematics, setting him on the path that would eventually lead to an undergraduate degree in engineering from the University of Manitoba. Gerry fast-tracked his undergrad degree by convincing administrators he would be better served taking more advanced courses in numerical analysis and other scientific subjects over the prerequisite undergraduate arts courses. Gerry soon became hooked on mathematics and modeling, developing the basis for his future research and the company's computerized production-control system that remains in place today in several of Price's North American factories. "That was my nature in engineering and science. I really understood the fundamental first principles of science. It's all I had to know. From there, a person can create any formula and derive anything."

While his studies were rigorous, the academic schedule was unrelenting. "The workload was huge. I put in very long hours. One of my absolute 'aces in the hole' was my steady girlfriend and love of my life, Barb. We started dating in high school and were together from that time on, all the way through university. I worked long hours, getting used to living on little sleep."

Gerry's endless energy

To understand how Gerry led Price through the tumultuous formative years of the 1980s and 1990s is to also understand what sustained his endless energy and ability to continue on when just about everyone around him told him he should give up. His ability to work through massive workloads took hold in his university years as an ambitious engineering student studying weather forecasting. His thesis work predicted what the atmosphere would do based on complex mathematical equations charted out on three-dimensional grids. "As a recent newlywed, I was in the process of completing my undergraduate thesis in engineering. I needed to work around the clock, so I got an air mattress and sleeping bag and literally slept on the floor of my office at the university," remembers Gerry. "It was not exactly romantic!"

Alberta bound and back

Ironically, at that time in his life, Gerry had no desire to become a businessman and eschewed the idea of a corporate career. "I had zero interest in business." He went off to Lehigh University in Bethlehem, Pennsylvania, to earn his PhD in engineering, thinking he would one day be a university professor. With his love for mathematics, he excelled at Lehigh, teaching undergraduate thermodynamics and earning a grant from the National Science Foundation for his work in atmospheric modeling.

Gerry's first official job out of school in the mid-1970s was working for the Government of Canada's Defence Research Establishment Suffield (DRES) facility at Canadian Forces Base (CFB) Suffield in Ralston, Alberta. As a defense scientific services officer, he leveraged his mathematics to predict the transient responses from blast waves on warship antennae, utilizing data from experiments done at the White Sands Missile Range in New Mexico. As Gerry explains, "The blast wave coming across the desert hit the antenna, and the antenna deforms but remains functional, or it gets sheared right off. One or the other happens, right? And my job was to predict that response mathematically. The experiments at White Sands Missile Range were to measure the response accurately and to see whether the model agreed with the measurement, which it did." Establishing successful modeling methods not only set up Gerry for success early on in his career but would also serve him well as he expanded Price into the US.

Adaptation over obstacles

One obstacle that Gerry has adapted to with resilience is his visual impediment. "I figured it out when I was in grade 5, playing baseball. I was trying to catch a fly ball. I got beaned in the head a number of times because I couldn't find the ball, but I didn't know why. Monocular vision was detected, noting that binocular vision is required for distance perception. My body only sees with one eye, my right one. My left eye works if I close my dominant eye, and then my left eye cuts in. But it's not really aligned with my dominant eye. My brain has suppressed that image. For my entire conscious life, I've known only the right side of my nose."

Whether developing strategies to successfully play sports in school or effectively locate certain people in an audience during a presentation, Gerry focuses on adapting to the opportunity—a precursor to situations that would test his resilience in the years ahead.

Gerry's "MBA"

Not having a formal business education or management background did not deter Gerry from building a successful international business. Instead, he relied on his engineering and mathematical background, where knowing how to solve problems, program computers, and create code served Price, helping to automate the plant in Winnipeg in the 1980s. "I knew computers from a scientific perspective, not from a business perspective," Gerry recollects. "I also relied on my mathematical training to develop my own profit-and-loss reporting systems, which I still rely on today."

That intuitiveness caught the attention of Gerry Law, then President of Price and whom Gerry had worked for as a summer student at the company, implementing computerized production-control processes in the factory. Law soon enticed Gerry away from his government job in defense research to Price, where Gerry eventually worked his way up to the role of president. This significant career risk of leaving his "secure" government job in Alberta was one he would draw strength from as he went on to take even greater risks, many of which made Price and its affiliate companies what they are today.

In the chapters ahead, you will read how the 13 Tenets that make up the Price Way served Gerry when he became the President of Price in the mid-1980s. "In each of those jobs, from general manager to division president, I learned what I would or wouldn't do in business. My first several years at Price was like my MBA."

 **Service**

GERRYISM

A life of service is a life well lived.
Everything we do is about service.

SERVICE TENET

When we serve our customers better than our competitors do, focusing on our customers' success rather than our own, our customers will thrive, grow, and ultimately lead in their marketplace—and so will we. At Price, service means having short lead times and being on time every time, providing quality, uniquely engineered products, and offering expert support so that it's easy for our customers to do business with us.

Service, it fits to a tee

"It was a white T-shirt. It had a blue rim around the neckline and a blue rim on the sleeves," remembers Arnhold Neufeld, Price Production Cell Leader, speaking of the first T-shirt he received from Gerry when he started at Price as an impressionable 19-year-old "kid."

Little did Neufeld know just how much of an impression Gerry's gesture would make. The T-shirt had an art deco graph that displayed the goal of decreasing lead times from five weeks down to three weeks, accompanied by a cartoon of a Price truck going down a "bumpy road" to symbolize that reaching the goal would not be easy but certainly was achievable. Lead times in the industry in the late 1980s were 12-plus weeks. Four decades after Gerry first placed that T-shirt in his hands, Neufeld still remembers that he knew immediately that the T-shirt was symbolic of something much bigger. "Gerry's serious about decreasing lead times, that as a business, we need to get there so we keep our jobs and stay competitive. I took that message to my own duties on the job—to get orders done correctly and efficiently."

The decreasing-lead-times T-shirt would be just one of several Neufeld received directly from Gerry over the years— each one illustrating an important operational goal. "He really got the message through to every staff member. And by the way, the company *did* decrease lead times from 12 weeks to 10 weeks, then from eight weeks to five weeks, and eventually down to three," Neufeld says proudly. "I'm kicking myself. How cool would it have been to have kept all those T-shirts over the years?" Neufeld asks, with a hearty laugh at the memory.

When Gerry became President of Price in 1986, the company's marketing material was designed by the engineering department. The content was accurate and functional, but Gerry felt it could also be creative. Leaning on Barb's design background, the company vetted candidates and hired a professional designer, Rick Nielsen, to design the T-shirts and create what would become Price's first marketing department. Gerry gave Nielsen the Price Way Tenets, which were about eight at the time, and Nielsen designed cartoons for the T-shirts.

While the T-shirt may be gone, the message is as prevalent as ever; keeping lead times down and fulfilling orders on time remain foundational strategies of Price's commitment to service.

During the April 2022 Bankers' Orientation, Gerry reinforced the importance of this goal. "We tried to take our lead times down to ridiculously low levels on products that no normal human beings should be able to produce," asserted Gerry. "We rolled out new products at an incredible pace through our research labs, and we refreshed legacy products continually." Looking at the room packed full of his closest supporters and stakeholders, he continued, "And lastly, we make it easy to do business with all our companies by having cutting-edge software to enable the ordering of products and having friendly people answer the phone, not a recording. Those three elements are what our service is all about."

For some orders, that dedication to service produces products that literally support life-and-death situations. For surgeons, where every second counts to save lives, Price's hospital operating room diffusers (Ultrasuite) give medical professionals pristine lighting conditions to better do their job plus a continuous supply of purified air at the surgical zone. Given these circumstances, the sense of urgency to get these products into hospitals can be demanding. From sales to design to production to delivery, fulfilling these orders on time underscores the value and importance of service.

From hospitals to corporate campuses to major developments around the world, Gerry says the one Tenet that underpins all others in the Price Way is *Service*. "Everything we do is about service. We just keep expanding the areas that we can serve, which is why we grow. It seems to me that it makes sense that if we keep our customers happy each year, and happier each following year, they're going to win more. And when they win more, guess who wins more? We follow them right up the curve because we ride their coattails to success, which is why we grow."

Service is simple, but it is not always easy

While a commitment to service may sound simple, it is not easy—a challenge Chuck Fraley, President of Price USA discovered early on in his leadership role. "When I first joined Gerry, we were relatively small, and sales were modest. Customers believed in him, but we weren't there yet. We still had to build the capacity and the capability to serve," says Fraley. Marking his 25th anniversary at Price in 2022, Fraley states, "The philosophy was there, but the mechanics of service took a little time. Eventually, the contrast between us and our competitors became evident. There was often commentary on how refreshing our commitment to service was. It was a big part of our growth. Customers knew they could depend on us. And that's it. It's so simple."

But remember, simple is not easy, which is why not everyone within the organization was on board with a service-oriented approach. "When I joined Gerry in 1997, we had a developing leadership team, but there were people on that team that were not as customer oriented as I was. And so there was some work to do, but what was delightful for me was that I had the support of Gerry. Some of those folks are no longer with us. It was a journey. But we found our way, bit by bit, and things got better and better over time."

Gerry assessed service was a missing key component from the air distribution segment of the US HVAC industry. He knew if he could instill the service-first culture in his American operations, it would give him the greatest chance to succeed in an ultracompetitive market, especially as a Canadian company that was a "nobody." He needed the right people to come on at the right time.

After retiring from one of the company's biggest competitors, Marv Dehart joined Price USA in August of 1999 as a consultant, eventually becoming President of Price USA in 2005. "Dehart was a proven expert in operations and lean business principles. He joined the company at the beginning of the service transformation," recalls Fraley. Fraley worked very closely with Gerry building Price's powerhouse sales representative network in the US for 18 years before succeeding Dehart as President in 2015.

Both Fraley and Dehart came from corporate backgrounds in the 1970s and 1980s when earnings came before just about everything else. "Marv got it. He took Gerry's mandate to heart, and he transformed the US operation with sales tripling from 2004 to 2008. We made a commitment to service, and that was the beginning of building what we have today," Fraley adds.

The Price was right

Gerry's determined yet humble approach, complemented by a vivid vision of what he was attempting to do with the US expansion and contrasted against a formal corporate culture, made joining Price USA an easy decision for Dehart. Making service a top priority was an attractive incentive to recruit and retain talented, top-tier executives like Dehart. Recruitment would be essential for Price's ambitious expansion plans stateside. According to Dehart, "It was very rewarding to come out of what I call 'working for corporate America,' where satisfying shareholders was generally the primary responsibility along with having a strong financial bottom line." Certainly, they are important components of a successful large-scale business, admits Dehart, but he and soon others were looking for more— to be of meaningful service. "Understanding Gerry's concepts and his way of doing business was really enlightening to me. I learned a new way of looking at a business, much more driven by service. I think service is number one. Gerry always talked about how service fuels growth, and I have to say that he was absolutely correct from that standpoint," says Dehart.

Return on service

A key performance indicator for Price coined from Gerry's service commitment is return on service (ROS), which is heavily referenced by Gerry and others when prioritizing investment in supporting the customer. ROS, above any other business strategy, correlates customer satisfaction to long-term sales for

Price. The COVID-19 pandemic put that premise to the extreme test. "I don't know of any other manufacturing company that did what we did during the pandemic," reveals Joe Cyr, President of Price in Canada.

Just days after the pandemic shut down manufacturing operations for almost every industry across North America, Price, led by Gerry, leaned into its *Service* Tenet at a time when it was seemingly impossible. Cyr remembers calling customers to "apologize" for delays in shipping Price's HVAC products; supply chains grinding to a halt because of uncertainty caused by the pandemic was "not an excuse"—a message Gerry instilled in leaders across all their companies.

Many success stories are the result of failure, and Gerry elaborates on why he felt it was important to apologize to clients. "We were too slow to see the reality of our inability to hire additional staff, too slow to see the reality that a surge was building that was damaging our ability to maintain our service promise. We failed." That staffing failure is an admission that comes as a surprise to many people close to Price, including leaders closest to Gerry. As the pandemic represented a juncture in the company's growth where innovation facilitated massive sales and service opportunities, Gerry indeed recognizes what went well while remaining humble regarding what did not, specifically hiring more people when it mattered most.

In interviews with Price leaders, common recollections of stoic and steady leadership are synonymous with how the company led with service during the pandemic. "As we fell into the abyss of the pandemic like everyone else, Gerry led from the front," says Cyr. "He went plant by plant and never missed a minute. It was clear from the onset that he expected us to use everything we had, including vertical integration, financial strength, sourcing power, ingenuity, sheer will, and effort to fulfill the service promises. We pulled out all the stops and eventually delivered on the service promise when most of our industry, and most of the business world, could not and did not."

In the waning months of the pandemic, Gerry, Cyr, and other Price leaders debriefed which processes worked and which could be improved. This analysis solidified the phrase *return on service*—a term now firmly ensconced into Price's corporate lexicon. The affirmation of the return-on-service commitment has transcended sales, fortifying client and partner relationships. Cyr states, "It was about them, not us. Our customers aren't just customers; they are committed partners because of what we did and what we do. This is all because of the service promise, Gerry's leadership and commitment to service, starting out with Arnhold's T-shirt."

Service is a journey, not a destination

With a service motivation prioritized over 25 years, Price progressed to become a "one-stop shop," making it easier for customers to find just about everything they needed in one place. One example is the movement of key services like anodizing from a third party to an in-house process. Anodizing is an electrochemical process that converts a metal surface into a finish that is extremely durable and resistant to corrosion, while at the same time decorative. "Historically, we sent out our material to a third party to have an anodized finish," notes Jeff Rogers, Director of Customer Support Services at Price's plant in

Suwanee, Georgia, adding that this in-house service makes Price more competitive as now the company controls the turnaround time from start to finish. "It gives us a competitive advantage by being able to offer faster lead times."

Consistent with other investments made by Price, the anodizing process was based solely on service. On paper, the payback period for this multimillion-dollar investment was almost 15 years, but the ROS was immediate, saving customers time and the expense of looking elsewhere, declares Cyr. "Having this service 'in-house' drew in a lot more business, so as is often the case, the payback was more like three years. As Gerry likes to say—build it, and they will come."

Price's service approach extends internally as well, to all companies under Anvil. Marty Maykut, President of Anvil, suggests the company's commitment to service not only serves its key stakeholders, but also influences the HVAC industry. Citing reducing lead times as an example, he says, "I think it actually shifted an entire industry; the industry might ship at eight weeks, and then Price would ship at week six or five—sometimes even less. You can change an industry simply by being a better service provider. I think that is truly serving your customers when you are providing something that others are not and you're doing it differently and in a better way."

Maykut, a practicing civil engineer who came to Anvil in 2015 after building his career in the condo-construction industry, shares candidly that he was hesitant to "join the family firm" as he is married to one of Gerry and Barb's two daughters, Shandis. While the girls (Ainsley and Shandis) have always remained active in Price Industries at various levels, they have chosen alternative career paths to Price leadership. After studying the company closely for 10 years, while attending company meetings and events, Maykut made the leap to leadership, working his way up the ranks to his current role at Anvil. "In my career prior to Anvil, I learned that delivering on your promises means everything; that's how we grew our business successfully. With a background in business, and a close tie to the Price family inner circle, Maykut offers, "I am as deeply committed to Gerry's service mantra and the Price Way Tenets as he is."

●

Price Points on Service

Embrace fast lead times
Price rolls out new products at an incredible pace through its research labs, while refreshing legacy products continually. Having cutting-edge software enables the quick ordering of products by customers.

Welcome human connections
Price has real people answering phones, not robotic recordings directing the customer to a corporate website.

Understand service is simple, but not always easy
Expect that not everyone within the organization will be on board with a service-oriented approach, at least not initially. Some people will get on board and others will need to leave the organization.

Create a culture of return on service

Difficult times, such as the pandemic, put corporate values to the test, but just as service has become fundamental to Price's success, so can it be for any company.

Take the opportunity to influence

Price's service approach has influenced many facets of the HVAC industry. Just simply by being a better service provider, you can show a better way to do business.

 # The Golden Rule

GERRYISM

Treat others as you would like to be treated yourself.

THE GOLDEN RULE TENET

"Treat others as you would like to be treated yourself."
Irrespective of the outcome, by living a principled life in
accordance with the Golden Rule, you'll leave a legacy that is
worthy of you, and you can celebrate because you did it the
right way. You can go through life taking or giving. We prefer
the giving approach—it's the only legacy that's worthy of a
life lived.

Never break the Golden Rule

The shouting of two senior leaders pierced across the office space, bouncing off the walls and ricocheting off the senses of anyone located on the second floor. As with most senior leadership teams, testy exchanges between Price leaders are sometimes heard, though ultimately they dissipate respectfully with the mutual understanding of the task at hand.

However, *this* exchange escalated where others faded, erupting into a full-tilt, hoarse-throated, accusation-laden, highly emotional diatribe of one senior leader toward his colleague, as he expressed his ego-infused value to the company. Heads turned and discussions hushed as the ranting leader stormed on, leaving the other leader stunned.

The opportunity to "seek to understand," as suggested by author Stephen Covey, passed, as the ranting senior leader doubled down on his vitriol, holding firm his value to the company. For the sake of playing the devil's advocate, the leader was indeed foundational and integral to the growth of Price by creating and implementing services, products, and operations that the HVAC air distribution leader has built its reputation upon. Nonetheless, the Golden Rule had been broken.

To succeed at Price, either internally as an employee or externally as a partner or supplier, one must strive to never break the Golden Rule.

When building a billion-dollar company, while concurrently becoming an industry leader, there are bound to be some stressful and intense business circumstances that can break down even the most unflappable leader. Everyone is human, after all. A great deal can be smoothed over and forgiven with the humility of "cooling down," a genuine reflection, an apology, and a commitment to do better. However, the offending leader never relented, apologized, or reflected. This contentious exchange was offered by Price leaders as an example of breaking the Price Way's *Golden Rule* Tenet.

Recalling that fateful event, Cyr retells what happened next. Heavy-hearted and deeply concerned, he approached Gerry on what to do. "He looked at me calmly, knowing the senior leader in question had pulled off miracles and was a rising star. Gerry listened to the story, staying composed and said, 'He's breached the Golden Rule in a way where there is no road back.'"

A week later, the unrepentant leader was no longer with the company. "There was no choice in the decision, no matter how good the person was at his job," asserts Gerry, admitting that following the Golden Rule in this circumstance was difficult, but unquestionably necessary. "I loved this person, and I worked with him so long and so well, but I didn't know about his dark side." Learning from the experience, Gerry has established checks and balances to consistently evaluate Price's leadership, wanting to avoid another circumstance that would require letting a senior leader go—but more importantly, ensuring no matter what your position is at Price, you work toward a healthy working environment. "What counts is how leaders treat others and, more importantly, how they treat the people in the business with the least authority and power. Anybody who abuses people or takes advantage of people who can't fight back because they don't have rank, privilege, or education ... I have zero tolerance for that—zero."

Don't break someone's rice bowl

"Don't break someone's rice bowl" is a proverb that has taken on many iterations over time, but the meaning generally remains the same: the bowl represents a person's ability to earn a living. Break someone's rice bowl and you risk irreparably damaging someone's livelihood. Over the decades, Gerry has faced some difficult decisions managing stakeholders, both internal and external, and if the infraction egregiously breaks the Golden Rule, the consistent outcome is to part ways amicably. Gerry took difficult action whenever the Golden Rule was broken, even parting ways with offending managers who were instrumental in the growth of the company. When "letting go" senior leaders with just cause, Gerry aims to never break anyone's rice bowl. "Take care of people, even if they've made a mess," states Gerry. "Even if it's an acrimonious parting of ways, take the higher road to allow them to hold their heads high and find their way forward. It's the right thing to do every single time."

Following the Golden Rule is more challenging when "nice people" are no longer succeeding at the company, something Gerry encountered when he first joined the company. "I had to let go a receptionist who was not expecting it," reflects Gerry. "That interaction was very difficult—teaching me to always remember the human cost of parting ways with staff. I always say, 'You know, it's not working out. It would be better for you to be at a place where you will be having more job satisfaction. Stay on board while you look for a job, as that's an option. Let's see what you can find in six months. It's always easier to get a job when you're employed.' In several cases, I kept them on until they could get a new job." That overture was extended as well to the senior leader described at the beginning of this chapter. "Although he chose to not take it, we offered to keep him completely intact, even though he had breached the Golden Rule," shares Cyr.

Following the Golden Rule can be expensive

The Golden Rule may sound simple in theory, but it is challenging in practice and can be expensive and time-consuming.

One of Price's biggest supporting infrastructure projects tested the company's commitment to the Golden Rule. "There was a situation where resources and supplies were going missing on-site as we embarked on supporting this gigantic project," remembers Cyr. "The challenge was the client was insisting it must be our fault, and we had to ship replacement product at our expense." That expense was running up a replacement-parts tab well into the millions of dollars, and millions more were being spent sending Price engineers to the site to provide support that was not included in the original contract. "And the Golden Rule meant putting ourselves in their shoes, knowing it'll straighten out in time, doing the right thing," assures Cyr. With additional on-site support, on Price's dime, the missing-product problem was ultimately figured out. "Eventually, they acknowledged that it was an error on their part. We didn't stand on principle, and we didn't hold them for ransom." Empowered with autonomy by Gerry to always find a fix, Cyr and his team knew the priority was to find a solution, seeing the extra, unexpected expenses as long-term investment.

"We just looked at it from their eyes. And when they needed help, we made it happen," insists Cyr. That long-term investment of unexpected resources and supplies, at Price's expense, paid off, strengthening trust with a major customer, giving the client the peace of mind that the Golden Rule was backed up by action.

Role reversal is fundamental to the Golden Rule

"We follow the Golden Rule," avows Gerry. "It basically means, in any situation, pretend you're in a role reversal and you're on the receiving end of what you're doing. And then ask yourself, 'How would I like to be treated?' and then you treat people that way. You would never build on the emotional fire," Gerry further explains. "When someone's angry at you, you never fuel the fire; you suck it up. You put out the fire, helping even if it's not your problem. This kind of approach to the Golden Rule and service vision really are what we're all about."

"Committing to the Golden Rule takes consistent discipline to keep emotions in check," says Chuck Fraley. As Price USA President, he ascended from a vice president of sales role, where he faced demanding sales situations where the customer was not always right. But instead of placing blame, he learned taking the high road paved the way to success in the long run, first by solving the problem and then second by fixing any miscommunications or missteps once cooler heads prevailed.

"Sometimes that meant we ate a little more than maybe we deserved," admits Fraley. "But at the end of the day, a win-win resolution was always the goal. I always remember Gerry teaching us you never win an argument with a customer, and he's right," shares Fraley, noting rationale for this book without prompting. "As we grow and get larger, I hope we're able to hang on to all that has helped us become who we are." That includes the Tenet of the *Golden Rule*, most notably in the throws of one of the most onerous of all business outcomes gone awry—lawsuits.

"We've been involved in a couple of lawsuits," reveals Fraley. "These people that come at us are bitter and they're angry, and they're spiteful and they want to hurt us. They want to extract as much value as they can and to be as disruptive to our operations as they can." Remembering a specific incident, Fraley noticeably shortens his breath, pauses to recall the intensity of the circumstance and then channels Gerry's Golden Rule pledge. "It would be easy to become emotional and responsive and do battle. Gerry pragmatically steps away to separate the emotion from the issue, challenging us to ask, 'What are we looking at? What is the best pathway out of this in the best interests of the business with the least amount of impact to us?'" reflects Fraley, admiring Gerry's ability to find mutually beneficial solutions. "And he does that again and again and again. He's got this amazing ability to detach himself emotionally from these, what otherwise could be very charged situations, and it has served us really well."

The Golden Rule's silver lining

At some point in time, despite being of utmost service to the customer, deals fall through, or partners act in bad faith. It is simply the nature of business. Yet what distinguishes the longevity of one business from another is how its leaders navigate the business through a rough sea of emotions when money and reputation are at stake. Understanding there is always a silver lining to the high cost of the Golden Rule is a lesson Gerry learned during his initial years as a vice president at Price. In spite of anticipating losses from the ceiling and interior-finishing contracting business that he had decided to wind down, Gerry completed every contract to conclusion, including a hospital project that resulted in a half-a-million-dollar loss. Even with the high cost to complete these contracts, it was the right thing to do.

For Price and Anvil, the Golden Rule is also a reflection of business acumen that embraces short-term pain for long-term gain. Reflected in the details, right down to the proverbial fine print, the Golden Rule is held sacred, enshrined in legal documents and agreements to protect Anvil's reputation and integrity. Drafting, reviewing, and overseeing many of those technical documents is Janet Racz, Anvil's Legal Counsel and Corporate Secretary. "Gerry is always thinking long term, which works well with our legal philosophy because we're here to protect the company long term, not short term," stresses Racz, thinking of when Price considers the variables of reevaluating or, when necessary, exiting agreements and partnerships. "Even if it's going to cost us a little bit right now, we're thinking longer term, and we're okay with the short-term pain that sometimes people don't like to go through for the long term."

Anvil does not negotiate contracts with the end in mind, preferring to build rewarding long-term partnerships that benefit all parties. "Quite honestly, I've never seen loyalty like Gerry has for the long term," offers Racz. "If there is any insinuation of partners that are noncompliant on an important aspect of our agreements, we'll reevaluate. However, if you're a good partner or distributor, we're not going to be partners with you for a year to make money and then go. We're going to strive to be partners forever." And staying "forever" partners with Anvil will always come down to one rule—the Golden Rule.

With operations around the world in 11 countries, the reciprocity of the Golden Rule is integral to developing partnerships. In one overseas venture, Maykut shares, "We found out that they were making some of our products without our approvals. They were claiming testing standards of things that hadn't been tested." This behavior flouts the Price Tenet of delivering the *Straight Goods*. "That was a royalty payment of about a quarter of a million US dollars annually," offers Maykut, restating Anvil's resolve to uphold the Golden Rule—even at the expense of an annual payment of $250,000. Anvil negotiated out of the partnership, leaving the money on the table rather than work with a firm that misrepresented the product. "I would say we had a moral and ethical choice. That's a pretty big financial one."

●

Price Points on the Golden Rule

Live the Golden Rule
To succeed at Price, Anvil, and its affiliate companies, either internally as an employee or externally as a partner or supplier, you must follow the Golden Rule, treating others as how you would like to be treated.

Don't break someone's rice bowl
When employees no longer are a good fit with the company, they must go on their way. However, during that process, it is prudent to ensure that their livelihood is not risked, as referenced by the Chinese proverb "Don't break someone's rice bowl." People can always find a job more easily if they currently are employed, so offer employees a transition time to seek work elsewhere while still under the company roof.

Practice role reversal

In any situation, pretend you're on the receiving end of what you're doing. Ask yourself, "How would I like to be treated?" and then treat people that way.

Recognize the silver lining of using the Golden Rule

What distinguishes the longevity of one business from another is how it navigates its business through a rough sea of emotions when money and reputation are at stake. Paying to exit a deal or business that could harm a company in the long run is worth the short-term pain and cost.

3 Straight Goods

GERRYISM

We guarantee the performance of every product we build and sell, and our customers trust this is the case and go to the bank on it.

STRAIGHT GOODS TENET

We design and build complex, engineered products, and we guarantee the performance of each one. Providing straight goods means forgoing gimmicks and never relaying misleading information. Our customers count on us to always give the "straight goods" in everything we say and do, whether it be product performance, application engineering advice, or delivery information.

The sound of success

Located in Winnipeg, Manitoba, Price Research Center North (PRCN) is a stunning 29,000-square-foot state-of-the-art laboratory and testing facility. The center consistently provides some of the most accurate and reliable test data in the HVAC industry, essential for producing products that deliver the straight goods. PRCN is a showcase facility for Price that serves as a benchmark for HVAC innovators around the world. In some foundational ways, PRCN in Winnipeg may just be the unofficial affirmation of Gerry's *Straight Goods* Tenet as its testing mechanisms and processes produce the accurate data that Price's customers have come to rely upon.

For example, you would think that Price's emergence into the noise control market and this remarkable research facility came together smoothly after a well-planned design followed by a streamlined process of development.

Nothing could be further from the truth.

In fact, Price's entry into the noise control market was an unceremonious financial free fall. Noise control products include panels, silencers, diffusers, and more, which dampen or even eliminate the sound from noisy HVAC mechanisms that resonates through the ductwork in non-residential buildings. The plan was to "start from scratch" after the original plan to purchase another facility fell through, recalls Cyr, exhaling, while shaking his head and smiling at how the projected investment to build the noise control test chamber ballooned from $750,000 to well over four times that price tag.

In 2004, Price wanted to enter the noise control market by purchasing another company, along with its research and facilities. "We knew noise control naturally fit our air distribution products," confirms Gerry. "Not only was it a natural kind of expansion of the business, but they also were complementary and reinforced our model. There happened to be another player in town already with a footprint and some knowledge, along with a lab, industry presence, and sales. Our first try was to see if we could acquire them, but they used us as a vehicle to sell to someone else with higher numbers," sums up Gerry, succinctly offering his assessment of the deal gone awry. "We were exploited."

Price is fiercely bullish when it comes to going out on its own, developing new products while intently innovating the research and technology at the same time. "We just reverted to doing it the way we've done everything else, which is to start from scratch," says Gerry, gaining confidence in every word on how he turns failure into success. "The pivot point was being duped and having failed. Okay, let's build on that." And that is exactly what Gerry and his team of engineers did in 2004—conceptualizing, researching, and building the $4 million test chamber at PRCN.

Paraphrasing the philosophical statement on staying optimistic, as one door on a noise control facility closes, another one opens, even if it does not exist yet and means building it from the ground up. "When we entered the market, we did it right, starting with the very best lab we could build. We leveraged our engineering know-how and intrepidness to figure it out from scratch," remembers Cyr, noting it took over three years before Price was introducing new noise control products into the market.

The truth triumphs in the end

By 2009, now five years into the research and development of the PRCN expansion, the business unit was bleeding millions of dollars per year due to the ambitious undertaking. Price was eager to start seeing a return on its bold investment. With the lab now in operation, pumping out engineered products backed by innovative science, the business should have been full steam ahead. It was on the right track, but with one major derailment.

"We were getting beat up by the leader in the market because they were publishing performance data that suggested their products outperform ours in terms of noise dissipation," defends Cyr, still frustrated as he remembers the "sleight of hand" of the competition. "The research team at PRCN knew something was amiss with the lab numbers provided by the competition. It was a head-scratcher. We hired a young engineer out of the acoustics industry from a large-scale consulting company in New York City. From his independent analysis, we discovered that our competitor was publishing results that were not reproducible in our lab," shares Cyr, underscoring that this tactic hurt not only Price but the entire HVAC industry.

"It would be wrong to say they were lying, but they were presenting data and research in a 'smoke-and-mirrors' fashion, using invalid assumptions," recalls Chris Dziedzic, General Manager of Price's Noise Control business in Winnipeg.

Gerry's Tenet of *Straight Goods* means sticking to the truth, no matter the challenges from the industry or the harmful tactics of some bad actors. In the long run, staying on the straight and narrow paid off for Price. In essence, sticking to the straight goods was the company's way of retaliating against unfair competitors, ethically done with irrefutable contrasting data that its customers soon learned to understand and, more importantly, trust. "We knew we were getting beat in an untoward way—so what to do?" Dziedzic questions. "We told the truth—but began to share data in two formats: the proper and valid way, which is our method consistent with proper engineering and science, and then the other way 'for information purposes,' using the same conditions and assumptions as the other guys." The response from customers seeing and understanding the contrasting data was immediate and impactful. "The benefit was that customers could see right through this. Not only did they know we were being honest, but they could see they had been misled by the other guy without us calling them out." Once Price amassed its own data, backed it by science, and packaged it alongside its competitor's "research," they armed the sales representatives to educate an industry. "We simply provided our reps two options," says Ted DeFehr, Vice President of Sales for Price USA. "We obviously found that when presenting our data based on the older 'standard,' our data matched up much better."

Furthermore, Price's training centers, product displays, and software tools provided the information its customers needed to select the most effective and efficient solution or product. With commercial-grade air distribution products, performance data is used by engineers to determine exactly how air will be distributed through the space. The data includes the pattern and shape the air takes, as well as the velocity of the air in cubic feet per minute (CFM) as it is distributed. For noise control products, the data relates to decibels (dBA) and sound curves in a frequency band heard by the human ear. Independent industry bodies in the

HVAC industry create test methods for air distribution products and update testing standards on a regular basis.

On the road to straight goods

"But how do you prove it works?" is a question Gerry asks himself relentlessly when assessing if the products his company produces are indeed living up to the *Straight Goods* Tenet. Beyond the Tenet, the answer to that question serves hundreds of sales reps that offer the products to customers around the world. "How do you prove it's cleaner air?" asks Gerry intently, looking off distantly to relive the experience deep within his always-in-motion stream of consciousness. "Well, the best way is to measure how your products perform, which we did."

In the summer of 2020, Price began the process of designing and building a lab chamber to track individual aerosol particles similar to those of the size of what would be expelled by a person's breath or cough. The sophisticated measuring tools and techniques permit engineers at Price to measure the concentration of aerosols from a simulated breathing individual. The first reason to measure the concentration of aerosols in such a scientific manner was to prove that the air purification products that Price manufactures actually work in the real world. The second reason to measure in this way was to compare lab-measured performance to predicted performance from the Price computational fluid dynamics (CFD) model. "We did that in collaboration with Purdue University's Dr. Chen, who is a world-renowned expert in CFD, credited with hundreds of publications in this area," claims Gerry. Price's research team and Dr. Chen's team compared CFD predictions with particle-tracking measurements in Price's research facilities equipped with and without company-produced air purifiers. The research was published in a peer-reviewed journal article called "Investigation of Airborne Particle Exposure in an Office with Mixing and Displacement Ventilation," which was published in the April 2022 issue of *Sustainable Cities and Society*. A second paper, "Reducing Airborne Particulates Using Displacement Ventilation," was published in the December 2022 issue of *ASHRAE Journal*.

Between 2002 and 2003, Gerry made multiple trips to Europe with his chief engineer, Alf Dyck. In Europe, there was a greater focus on sustainability and the environment and the use of systems requiring less energy consumption, such as radiant panels, chilled beams, and displacement ventilation utilizing stratified air. In other words, the European technology was better at managing stratified air (cool, thinner air that sinks and warmer air that rises, balanced for greater efficiency and air quality) in commercial buildings. A hydronic system, also used in buildings throughout Europe at that time, heats or cools water and distributes it through pipes within the HVAC products, distinct from blowing warm or cool air at high velocities.

"Europeans had different practices than North Americans in this regard," advises Gerry. The goal of his visits was to learn from Europe's best practices. "I made a few trips to the big air-conditioning and ventilation show in Germany with Alf Dyck to learn firsthand what these European HVAC businesses were doing in terms of air distribution and comfort technologies. That's where we learned about hydronic and stratified systems."

Getting the facts accurate is one facet of the *Straight Goods* Tenet; another is corroborating the data through the knowledge of reputable HVAC innovators.

"You have to believe your product has value to the end customer. They may not see the value immediately, but you have to believe in your product. And you have to understand the value points in the product and be able to explain it in solid facts as compared to on the fly," itemizes Gerry, breaking it down into simpler terms. "This is a professionally engineered product that is actually superior to the alternatives that they're considering for their job." At the heart of the *Straight Goods* Tenet is establishing unquestionable research and data that confirm how Price products work, eliminating any doubt and fortifying confidence in Price's products.

"It has been the way we've been doing things for years. It's to inform customers of the options they could choose—pros and cons, helping them choose what is best for them in the long term. If it's not sellable based on HVAC fundamentals, then you're like a snake-oil salesman, pretending that snake oil cures all ills. But that's not selling; that's misleading and lying. Our job is to truly be able to communicate the points on why our products are superior," advocates Gerry, confidently defending his track record.

Telling the truth takes time and money

Sometimes you have to spend money to make money, and sometimes you have to relinquish the pursuit of money to make money—a steady strategy Gerry has lived by. "It costs a lot of money to tell the truth," confesses Gerry, noting the investment in truthfulness has always paid off in the end for Price. "Winning at all costs and compromising values, that's never been the way we operate. The key for us is how we play the game, not whether we win or lose. We will always play the game our way, and maybe we will win or maybe we won't. But what will never change is how we operate."

If anyone can attest to Gerry's dedication to the *Straight Goods* Tenet over the decades, it is Alf Dyck, retired Vice President of Product Engineering at Price. Starting in 1974, and retiring 44 years later, Dyck was Gerry's most trusted resource in developing many of the air distribution products Price built its reputation upon. "He was my barometer," admires Gerry. As for Dyck's recollections, he appreciated Gerry's trust and Price's resolute determination to provide the straight goods, recalling early on how the lab served to suss out the true science behind the products it manufactured under license when Price represented other companies in the HVAC industry.

Price worked with European manufacturers under license to design and produce products for the North American market. Inevitably, these licensing agreements had a limited shelf life as the pace of execution and innovation required by Price to fulfill its service commitment made them unsustainable. "We always honored our licensing agreements to the letter," reinforces Cyr. "It just wasn't a viable long-term service strategy."

In the meantime, while Price was under license with various products, its commitment to the straight goods meant rechecking and retesting all the performance data these manufacturers provided, ensuring these products, often redesigned by Price for North American use, performed as the data promised.

"Part of the PCRN's function was to make sure that what these manufacturers were telling us was totally accurate," remembers Dyck. "And often the case, that wasn't so, and we had to develop our own performance data. That kind of taught us a lesson that you can't really depend on all the information that's out there."

Dyck has directly influenced countless contributions to product development, all backed by science, leaving him little doubt about the return on investment and return on service. "That has helped the growth of the company; I think it's the main contributor to the growth."

Just because Price was producing the science-backed data did not mean its customers immediately accepted it, recalls Gerry. Price's commitment to the *Straight Goods* Tenet delayed the company's goal of becoming a leader in HVAC air distribution. "We clarified for the industry what the true numbers were. It took a long time for the industry to believe us."

Reengineering the straight goods

Appreciating the formative years of the Price Way's commitment to the *Straight Goods* Tenet hearkens back to the late 1980s. In 1989, the company's expansion into the United States was growing in earnest, with 25 sales reps in various regions. Keeping in mind that the US HVAC industry is gigantic, estimated at billions in revenue in the late 1990s, Price was therefore modestly in the US game, selling about $2 million on its products made in Canada to the US. Gerry was not celebrating that success for long. "It became very evident that it wasn't sustainable. We knew it wouldn't be sustainable, but it allowed us to build relationships. It allowed us to fulfill some orders in the US and to start to get some incremental business. And most importantly, we started to realize where we had deficiencies in our product," analyzes Gerry, with his ever-present problem-solving mindset. "We had deficiencies in every single product we built. Every Canadian product was overengineered for the US market: too much quality, too much metal to perfect, and uncompetitive by anywhere from 20% to 50%."

Making matters worse, Price was aiming at a moving target. A competitor had become the leader in the commercial air distribution market through value engineering and lean manufacturing, making it harder for Price to match its standards. "Just as we thought we were getting competitive, they would strip the cost out of these high-volume products and value engineer them down to crazy low numbers. They were great at figuring out how to make these core, high-volume products for less money, using better tooling, equipment, and lean techniques," says Gerry.

Facing the almost unsurmountable challenge of reengineering products at a significant cost, Gerry explains how Price went through the painstaking and grueling effort of reworking its products, including grilles, registers, diffusers, and more, some taking up to two to three years before rereleasing back into the market. "One by one," recounts Gerry. "You can't do them all simultaneously, as back then, we never had enough resources."

The straight goods—now and forever

At the time of publishing *I'm Just Gerry*, Price was commemorating 75 years in business, taking the appropriate time to celebrate this rare accomplishment for any business and embracing the humility that all good fortunes in business can dissipate in far less time. Tasked as one of the leaders to guide Price into the *next* 75 years is Maykut. "I've always viewed knowing your current reality and straight goods as the same thing," offers Maykut on his take on the *Straight Goods* Tenet. "Don't sugarcoat it; know exactly where you stand. You need to know where you're failing so you can improve." Maykut maintains the Price Way's *Straight Goods* is as good as it gets in building a "forever company," sharing the Tenet will continue to serve Price and Anvil as a standard for now and for the future.

●

Price Points on the Straight Goods

Start from scratch
Price leveraged its engineering know-how and intrepidness to figure out how to enter the noise control market within the HVAC industry by starting from scratch. Patience and persistence are necessities when committing to the straight goods.

Triumph with the truth
The *Straight Goods* Tenet means sticking to the truth, no matter the challenges from the industry or the harmful tactics of some bad actors. Staying on the straight and narrow paid off for Price.

Do it with data
Once Price solidified its own data, backed by science, and packaged it alongside its competitor's "research," the company armed the sales representatives to educate an industry.

Tell the truth, even if it is expensive

Price invests money into research backed by science, noting the strategy has always paid off in the end. At the heart of the *Straight Goods* Tenet is establishing unquestionable research and data that eliminate any doubt and fortify confidence in Price's products.

 **Growth**

GERRYISM

Today a rooster, tomorrow a feather duster. Failure to adopt a growth business model virtually assures that any "rooster" of today is destined to be a "feather duster" of tomorrow.

GROWTH TENET

Growth is an imperative. Adopting a strategy of growth and scaling our business comes with risk, but both our successes and failures prepare us for even greater challenges in the future. At Price, growth is not something that happens to us; it is something we choose.

Gerry's blueberry pie

Like an ant on a blueberry pie. In the 1980s, that "ant" was Gerry and that "blueberry pie" was the United States of America. Many companies have tried and failed to do a start-up in the US market. All Gerry wanted was a piece of the blueberry pie—to become a leader in the air distribution segment of the commercial HVAC industry. At the time of publishing this book, Gerry's blueberry pie adventure is an undisputed success story, serving up a gigantic piece of that proverbial pie. As a result of Gerry's ambitious US growth plan, Price enjoys the number-one market-share position in the air distribution segment of the commercial HVAC industry in the US, setting the company up for enduring growth for the next 75 years.

As the saying goes, hindsight is 20/20. During Gerry's tumultuous tenure during those formative US-expansion years, many senior leaders, investors, bankers, and important stakeholders understood but did not support his lofty long-term vision, cashing out their investments in Price. Understandably, one can be forgiven for bailing out early, as many of the dire predictions for Price in the 1980s and stretching into the 1990s came to fruition, including greater-than-expected losses. Looking back, Gerry jokes now, "If I had a boss during those years, I would have been fired every year up until 1997 as my profitability predictions were off so dramatically."

"Not a single employee in the company wanted to do it," says Gerry of the expansion into the US market. "Not a single shareholder either. No one. They thought I was nuts."

If the first few years of the US expansion were any

indication, it would be easy to think that the nonbelievers might have been right. "I found it took about three times longer to get to breakeven. And the losses were more than three times bigger than I thought they would be," recalculates Gerry, adding up the cost mentally in his head. "In essence, the entire US start-up cost about 10 times more than I anticipated." Gerry estimates the first few years in the US put Price in the red by just over $8 million, which had an impact everywhere. "The total company was hanging on by a thread, year after year after year, throughout the entire decade, almost until the late 1990s," confirms Gerry, reflecting with a knowing nod, pausing, and leaning in as if to emphasize the following: "But I saw hope." However, hope is not a good business strategy, especially when putting everything on the line, and the reason Gerry had hope was because he saw the trends—and they were all pointing to profit.

To this day, Gerry harbors no resentment toward those who didn't believe in his vision, understanding the US growth plan was not for everyone. "Why not monetize it and cash out and not suffer the consequences? I was so happy to have the opportunity to take a shot at the US market. I was just ecstatic that we had the chance."

The blueberry pie was ready for serving

At the time of Gerry's entry into the US market, the air distribution segment of the HVAC industry was amid a seismic change; massive HVAC-related companies were amalgamating and, in the process, reducing or eliminating their capacity to serve customers. One company absorbing another was on trend in the 1980s where it seemed profits before people—specifically customers—was priority number one. At about this time, Fraley was recruited into Price, first as a US sales vice president and eventually becoming President of Price USA. "I knew that a lot of representatives in the market were getting frustrated with these other companies I had been with previously," shares Fraley. His intuition to leave a leading US HVAC competitor to join Price was quickly validated. "It was becoming more corporate. Price's competitors were less responsive, and the focus was becoming more operational than serving and supporting customers." Fraley's insight led him to believe that the customer-service gap was Price's market advantage. "The good news? We had this amazing owner who saw the opportunity and was willing to do whatever it took to win over these potential customers."

Knowing of the opportunity to grow the customer base for Price in the competitive US market was one thing; putting a strategy into motion was another. Starting with no customers in an ultracompetitive market, Gerry, along with Fraley, determined the best approach was to avoid the bigger US markets dominated by the company's leading competitors—at least, for now. In typical Gerryism fashion, Gerry even had a colloquial reference for this sales approach, calling it his hinterland strategy. Most dictionaries reference *hinterland* as a remote area far from major urban centers. "We decided which markets were going to be the next priority, and we would pick three," explains Fraley. As a new player in the market, Fraley says Price had the one ultimate selling tool others did not—Gerry.

"And granted, things would happen dynamically. We might not get all three or we might get two others we weren't even thinking about, but our goal was to try to bring on three significant markets each year. I would get into a targeted market and talk to the rep to understand what their needs are, and I could always count on Gerry to influence the sale."

An adept salesman, Gerry relished the challenge of turning people around who originally said no, finding pleasure in the art and science of selling by understanding who people are and what products they really needed. Gerry's introduction to the art of persuasion came when he first joined Price in the 1980s, securing a contract that was originally determined to be out of reach. "There was a job we went for in Saskatoon, Saskatchewan, and I thought we had the better product than our competitor. I was hoping we would get the order," recalls Gerry. "However, I read in the *Western Construction News* that the order was going to be awarded to a competitor." Undeterred and feeling emboldened by his straight goods belief that Price had the better product, Gerry called the decision-maker at the government ministry overseeing the project. "I said, 'I think you're going to make a mistake because we have the better product.'" As it turned out, the deal was not signed, and Gerry was granted a meeting with the ministry official. "We loaded all our gear into a station wagon, and the team drove it up to Saskatoon. We did a side-by-side comparison of our product with our competitor's and won the order," beams Gerry. Securing that contract was Gerry's first foray into the fabled "snatching victory from the jaws of defeat," and he says that lesson has stayed with him forever. "It's an interesting example of not quitting because there is always a chance."

Gerry's persistence paid off in Saskatoon, and he would need that experience as Price expanded into the US—to persuade sales reps to take on the company's line of HVAC products. As mentioned in the Introduction, a sales representative, or "rep," is not just one person but actually a team of sales professionals working for a sales representative company. Since Price was a new player in the US HVAC market, converting the US reps was a tough sell, Fraley admits. "If I wanted them to take the leap, I needed to get them in front of Gerry," says Fraley proudly, evidenced by his elevated energy in his smile. "And that's what I would do. If I couldn't get them to Price in Atlanta, we would go see them in their city." From his own experience of being successfully recruited away from a main competitor, Fraley knew the power of the Price story, delivered with passion and purpose by Gerry. "By the time we were done, they would be favorable to doing business with Price over a long-established competitor."

Getting a new client to shift loyalty from another HVAC supplier could take up to two years, though Fraley reaffirms what mattered most in a rep switching was not just the sales; it was the positive shift in sentiment toward Price's customer-service approach. "It seems so simple, and yet I marvel at how some other companies don't seem to get it," offers Fraley, "We were extremely customer-centric." Simply put, it was how Price grew in the US market—prioritizing one customer at a time.

Price's first $1 million rep

Building Price and its affiliate companies into a billion-dollar company took time and partners. Few of Gerry's partners have had the direct impact on Price's steady upward trajectory as the independent US sales representatives he has worked with over the years. As of 2023, the company does business with 60 sales representative firms, covering all regions of the United States. In Canada, Price operates its own distribution network through 18

E.H. Price branch offices across Canada with over 200 employees. In the US, Price's influence and power are driven largely by the robust network of sales rep firms that independently champion and sell products produced by the HVAC company.

It is hard to believe there was a time when Gerry could barely get the attention of just one sales representative.

Sitting in his office in Lenexa, Kansas, surrounded by family photos, awards, and mementos commemorating a successful career in sales, Kevin Harre is CEO and owner of Jorban-Riscoe & Associates (JRA). Over the years, Harre has made some bold decisions and has taken some risks of his own. Taking a chance on Gerry, who was completely unknown to Harre at the time, was not one of them. Receiving a business tip about a Canadian operation that could potentially fulfill some difficult orders that could not be fulfilled by his current supplier, Harre reached out to Gerry in 1989, who in turn flew out to Kansas from Winnipeg.

"I'll never forget the first time I met Gerry Price," remembers Harre, echoing a familiar statement that is almost as commonplace as someone asking, *Where were you when?* "He sits down and he asks, 'How do I get some of your business? We're not doing much with you guys.'" New to the drive that is synonymous with Gerry, Harre recalls being slightly but pleasantly taken aback by someone who was so keenly interested in being of service. "I grabbed five or six jobs that we hadn't submitted on, and I said, 'Each one of these jobs, for some reason, we can't fulfill.'" Harre's problem was the current suppliers he represented, which were also Gerry's competitors, were not able to provide all the necessary products to make the orders complete. It was almost as if Gerry was banking on Harre's customer-service gap. He asked Harre outright: if Price could make the products Harre needed, would Price get the contracts? Harre agreed immediately to award Price the series of contracts, amounting to about $100,000—a mere fraction of the millions of dollars in orders JRA manages for Price today. "Gerry made all those things happen, and from then on, the majority of our work was done with Price," exclaims Harre.

Building an empire, one rep at a time

Starting with JRA, Gerry and his US team slowly, but methodically, targeted other reps who serviced Price's bigger competitors, fulfilling orders by making products others could not. To this day, Harre does not know how Gerry designed and produced the 19 products that others couldn't. And as sales reps are well known to talk to each other, word got around quickly that Price could create new products for other reps facing the same gaps. According to Gerry, the "secret" to creating the new products where his competitors could not came down to the experienced team of engineers and draftspeople at Price willing to take on the foundational task. Primarily, JRA needed different types of air distribution products; Gerry's team carefully looked at each intricate design manually, designing and engineering to the specifics of what JRA needed.

"How they did some of those things was amazing," marvels Harre. With each contract that Price was able to fulfill where a competitor could not, the company created new customers who were eager to purchase more product from Price. Soon, Gerry—a virtual unknown—went to well known, says Harre. And before long, the strategy to influence smaller reps that were working smaller lines of HVAC products in the US grew to working with larger reps and larger product lines.

At the Bankers' Orientation in April 2022, Gerry shared just how successful the sales strategy has become when referring to a former air distribution market leader: "It's nice to double them. I would prefer to triple them," attests Gerry playfully, yet seriously at the same time.

Riding their coattails

In the chapter on the Golden Rule, Gerry professed if Price provided exemplary service to its customers and its customers succeeded as a result, the company would ride its customers' coattails to success. The same strategy applies to the sales reps that have helped Price grow to become a US HVAC industry leader in the air distribution category. "Our growth is paralleled with Gerry," expresses Harre. "We've had to grow our businesses to keep up with their growth."

As for JRA, the rep company is a good example of a small firm in a second-tier territory that saw the potential in Price, declares Fraley. "They grew with us over the years, and now they're the market leader in that territory today. In those early days, it was all about developing partnerships around a mutual desire to win and grow." Interestingly, over the next two decades of Price's phenomenal growth, many of the originating partner reps could not keep pace. As loyal as Price remained to its founding US reps, tough decisions needed to be made. "We would find it necessary to move up to a stronger partner," advises Fraley.

Price's growth correlates to its symbiotic relationship with its US sales reps, recognizing their commitment with the Price Cup. This annual award acknowledges a business partner that embraces values, principles, and operating disciplines that align with Price. Rist & Associates, the rep from Des Moines, Iowa, was the first to receive the Price Cup in 1997 and remains Price's rep in Iowa today.

Finding an oven

If the blueberry pie represents the US market, and a medium-sized piece of the blueberry pie represents the air distribution segment of the HVAC industry in the US, then Gerry needed an oven to bake his air distribution products—a US factory. Without one, Price could not sell competitively from Canada indefinitely as the Canadian market was contracting with fewer non-residential developments across the country. As documented in chapters throughout *I'm Just Gerry*, Price boasts some of the most technologically advanced HVAC production, research, and supporting facilities in the world, but that certainly was not the case in the beginning.

Wanting to get a foothold into the US, Gerry leased a dilapidated and aging plant in Norcross, Georgia, that was failing, a facility that had been previously used by a US competitor and before that by a German competitor that exited the States after an unsuccessful US expansion. "It was a pivotal point. Along came this failing plant in Atlanta that was around 80,000 square feet," remembers Gerry fondly, smiling at how humble a beginning it was for Price's US expansion. "It had a functional paint system and lighting, but totally antiquated and inadequate for scale—but enough to get a start with their workforce that supported the plant, and it was in the right city."

From 1987 to 1989, Gerry and his US leadership team scouted out factory sites in the United States before whittling the list down to three cities: Lincoln, Nebraska; Sioux Falls, South Dakota; and Atlanta, Georgia. Gerry learned from a previous plant start-up that it is better to rent your first factory before you buy one. That rent-before-buy strategy served Price well in 1989, giving the company valuable insight into working in the US before making substantial investments in infrastructure. "Norcross was a rented footprint with used assets that were second rate but enough to get going in the US market. It was a slam dunk. We purchased the used equipment at a huge discount, at roughly 25 cents on the dollar, and we picked up the workforce," highlights Gerry. "In any start-up, you can't afford to hire the best people at top dollar. As your start-up grows, top talent will come in time. In the meantime, recognize that everyone must wear multiple hats and complete numerous tasks that may be outside their skillset."

Serendipitously, the former Norcross plant owners were also building HVAC products as a licensee, strategically giving Price incremental revenue and immediate access to sales reps. The plant showed its age but gave Gerry the oven to bake his blueberry pie, producing products in the US market. More importantly, the Norcross plant gave Gerry the time to explore a bigger and better location.

Once the US start-up showed signs of being close to breakeven, land was purchased in Suwanee, Georgia, a suburb of Atlanta, and the first "owned" Price factory in the US was built in 1995. While modest in comparison to Price's modern-day facilities, the new Suwanee plant became the blueprint for all future US factories. Maykut cites the new aluminum-extrusion plant in Phoenix, Arizona, as one example. "If we're going to spend $130 million on a 300,000-square-foot facility and we're buying one press and we have room for three, let's spend an extra 10% and buy the second press," figures Maykut, explaining the extra investment as a precursor of better things to come. "Let's be optimistic that we can fill it up. Build it, and they will come. Spending a few million dollars a couple of years before you need it ends up looking pretty smart."

●

Price Points on Growth

Start small

Initially, in an approach introduced earlier in *I'm Just Gerry* called the hinterland strategy, Gerry targeted smaller sales representative companies to take on Price products. As the company grew in influence, Price went after larger markets and sales representatives. "The goal was to win but do so in a way that didn't draw attention from our bigger competitors who would crush us if they got wind of our steady growth in smaller markets," cautions Gerry.

Win with your partners

If Price's customers succeeded, the company would ride its customers' coattails to success. The same strategy applied to the sales reps that helped Price grow to become a US HVAC industry leader in the air distribution category. As Price grew, sales reps had to grow their businesses to keep up.

Find an oven

To grow, Price needed a US factory. Without one, Price could not sell competitively from Canada indefinitely as non-residential developments were drying up in the region. Eager to get a foothold into the US, Gerry rented a dilapidated and aging plant in Norcross, Georgia, and after some lean years, Gerry and his US team used this foothold to eventually build one of its signature plants in Suwanee, Georgia.

Grow for the future

Develop a growth philosophy, become a forever company. Start building the foundation today for growth 10 years from now.

 # Innovation

GERRYISM

The power to fulfill our dreams is within each of us.

INNOVATION TENET

We will never hesitate to proceed with a new business unit, investigate a new technology, or learn a new skillset if doing so will better serve our customers. Our company's accomplishments to date and capabilities going forward are derived from our continually acquiring knowledge and fortifying our technical foundation.

An overnight success story 11 years in the making

After deciding to abandon efforts to enter the precision-controls segment (patient isolation rooms, laboratories, and clean rooms) by way of acquiring a company in that market segment, Gerry asked his team if Price could instead build a business in this arena from the ground up. At the time of Gerry's request, Jarvis Penner, then a young Price staffer, raised his hand, signaling that he would like to assemble a team to give it a shot.

In 2011, Price launched Antec Controls, a division that manufactures air-pressure monitors and controls, targeting the multimillion-dollar lab, patient isolation room, and clean room segment of the air distribution market. Antec Controls' focus was to provide the critical research data Penner would need to build Antec. Two years later, Penner (who became Antec's General Manager within that time frame) and his team launched a series of innovative products that ensure the safety of healthcare workers, patients, and researchers by keeping critical spaces pressurized with pristine airflow—or so they thought. "We realized what we had launched in 2013 was good, but it certainly didn't hit the mark for where we thought we could be," admits Penner with humility. "And in 2016, we went back to the drawing board and started the development of what would be our second-generation line of controls, which would effectively obsolete everything that we had developed prior."

That which does not defeat you only makes you stronger, as the old adage goes—or at least, motivates you to innovate your products to make you an industry leader. "We learned a lot in those first five years about what worked and what didn't work," recalls Penner, adding that the opportunity to go back to the drawing board has only made the products better in the long run. "It was actually quite motivating because we could take all the lessons we gathered from that first seven years and apply them to make something amazing. And the product that we launched at that point, which we're still using today, is just remarkable." What is also remarkable is Price's commitment to innovation over the long term, considering it was a company with no prior history in critical-space controls and systems. In 2023, after posting straight losses for several years after start-up, Antec will realize sales in excess of $20 million.

"The difference between our competitors' thinking and ours is that as soon as we see a better way of doing something, even though we might have invested several million dollars into something to get it to that point, if we realize there's a better way, we start an investment program on the better way—right away," insists Gerry. Investment in innovation requires the discipline of putting product performance before profits, and Antec is one example of an overnight success story—11 years in the making.

All the right parts

Price made a bold move in 1999 to become its own electronics supplier, most notably developing its own firmware (the software that is hard-coded into a device, essentially telling the board how to function) and circuit boards. The investment came one year after purchasing circuit boards from an outside firm. Today, Price Electronics supports key products produced by Price, including variable air volume diffusers, air terminals, and the precision room-pressure control products of Antec.

The investment to create Antec was not taken lightly, even by

Gerry, who does not normally hesitate to jump in. "I was slow to the punch on this as an opportunity," divulges Gerry. After all, designing circuit boards for monitoring systems in critical spaces such as hospital operating rooms was literally a life-and-death operation.

In 1999, Price was producing a new line of smart ceiling diffusers, which are known as variable air volume (VAV) diffusers. A diffuser typically covers an air vent, distributing air throughout a room in a uniform flow. Price needed complex circuit boards to electronically control the flow of air right at the diffusers, which would provide individualized comfort in a space. "We found we were vulnerable to the rapid pace of the design cycle and having to buy circuit boards in volume that we'd then have to discard when we changed the design. If we could build them ourselves, we could control our destiny, reduce our dependence on third-party suppliers with long lead times and poor service, have no shelf obsolescence and in turn, serve our customers better. So away we went, learning on the fly."

Just about every electronic device these days has some sort of circuit board. With increased demand for circuit boards worldwide, Price instinctively knew creating these vital electronics in-house would reduce its reliance on suppliers and increase its ability to serve with short lead times. "Our products needed circuit boards and controls we sourced from others. We found the need to order large quantities from third-party suppliers due to their long lead times," evaluates Gerry. "As a result, we suffered from shelf obsolescence when we'd make a design change."

The increasing lead times for supplies ran counter to Price's lower-lead-time-and-on-time culture, and it was a conundrum that needed an imminent solution. Enter Mike Nicholson, joining Price in 1999, now serving as General Manager of Price Electronics. Nicholson, a young electronics technician, did it all—from hole placement of components on circuit boards the "old school" way to designing complex circuit boards and building a world-class electronics-manufacturing business. Today, Nicholson chuckles when he compares his humble beginnings as a team of one, working under a staircase, to today's team of over 120 electronic technicians and product designers.

The initial plan was for Price to purchase a company that was already manufacturing room-pressure control products. Nicholson remembers that evaluation trip with Penner to review the operation. "Gerry asked us if their products were show ready, and we told him that every one of their products will require some degree of reengineering. Gerry was very hesitant, considering the life-safety nature of the industry. That's when we decided, instead of buying them for their outrageous price for products we would need to reengineer, why don't we just take a stab at doing it ourselves? Today, Price Electronics serves a variety of customers and industries, including agriculture, healthcare, energy, transportation, and communications, with Price as its biggest client at 60% of its business.

According to Cyr, Price's Canadian President, without Price's relentless commitment to innovation, there would be no Price Electronics, and without Price Electronics, there would be no Antec. "They're a juggernaut, and they're a powerhouse," praises Cyr, noting Price Electronics' sales of over $25 million annually, with four high-speed robotic pick-and-place lines for high-volume manufacturing. Eventually, Price grew comfortable and confident that it could take on the critical controls market, and Antec Controls was born—with controls by Price Electronics. "Antec is now a thriving leader in the critical controls space, growing worldwide from a standing start in 2010."

Strong sales aside, Price Electronics mitigates many supply issues as the company is no longer entirely at the mercy of foreign-owned original equipment manufacturers (OEMs) overseas, celebrates Nicholson. "That's a key component of our vertical distribution and integration." That vertical integration became critically important during the COVID-19 pandemic. Fraley recalls, "Our integration really paid off in a pandemic where the world was crazy and you couldn't get stuff. It also helped us build our external market and helped us be stronger as a manufacturer."

During the pandemic, lead times for parts skyrocketed from 12 weeks up to 50 weeks, but as a result of the innovation to become its own parts supplier, Price Electronics had already stocked up considerably on raw material. "We didn't shut anyone down during the entire pandemic," confirms Nicholson proudly, nodding with affirmation on how early investment in innovation saved the day.

Looking back, Gerry credits the innovation of such business units as Price Electronics and Antec in establishing the company as an HVAC leader in air distribution. "These and other fledgling business units launched in the 2000s really differentiated us from the competition and enabled us to grow substantially, because once they became viable and started to scale our business, we went way up the food chain."

Slow and steady wins the race

Over the past four decades, a pattern of investment in innovation is evident: return on investment is a long haul at Price. According to numerous Price leaders, unlike most publicly traded companies and many larger private companies, Price does not fixate—or even calculate—investment payback periods on new ventures. As it relates to Antec, for example, after 12 years and $8 million of investment, the company is posting significant sales. "All we accomplished in the first 11 years of this process was to recoup our start-up losses. We're now at a net-zero basis," sums up Gerry, adding that the growth and profits on sales going forward look promising. The ROI realization is the same for Price's Noise Control division. "Nearly 20 years later, we have accumulated net profits to cover the losses on those first 18 years. The important lesson learned here that has carried forward in everything we do at Price is to invest in innovation to leverage our ability to drive incremental margin through products, just like we did in Noise Control."

Historically, on average, Price covers losses for start-ups for five to 15 years before seeing an ROI, underscoring an unwavering commitment to innovation. "Getting the science right takes time. Learning how to build it takes time, and marketing it takes time," justifies Gerry, his head nodding with the confidence that slow and steady wins the race. "Proper innovation cannot be rushed along. It takes what it takes."

The art of science and training

As the Price Way demonstrates, getting the science right takes time, and it also takes resources and training. Price built its own labs to get that science right. "If we didn't have our own lab, there's no way we could have split from a major competitor," asserts Gerry. "We built labs, category by category, to get the science right to develop the products."

Once the science is sound and the products are perfected, training is the next essential step. No different than its innovative approach to products and research, Price has invested over $10 million to build the Price Technical Centers (PTCs). Located in Suwanee, Georgia, and Casa Grande (near Phoenix) Arizona, these state-of-the-art technical facilities provide educational and informational experiences for customers, suppliers, sales reps, engineers, and other key stakeholders. The PTCs represent real-world conditions, featuring more than a dozen different product and application spaces, including visualization rooms that use theatrical smoke. The interactive space demonstrates Price's products under real-world conditions with hands-on training. "Engineers flock to us for our advanced levels of training," professes Gerry, adding visitors' time and investment are maximized by design. "They get continuing education credits when they get trained in our facilities." Developing, producing, and marketing products are fundamentals of Price's innovation strategy, but Gerry advocates that the strategy only works if customers know what to do with the product. "If we're going to be number one in the States, or anywhere, we needed a tech center to train the engineering fundamentals of all the products we have."

Supporting the tech centers is Price's product catalog. First requested at a Rep Council meeting in the early 1990s for use by engineers and sales representatives, the catalog has evolved over nine editions. Reflecting on its early distribution, Gerry remarks, "It took us to another level of perceived competence relative to our competitors in the market." Now over 30 years later, the catalog is 1,500 pages and weighs eight pounds.

"Anybody want to do exercises with it?" jokes Gerry in front of an audience of amused stakeholders at a recent Bankers' Orientation in Winnipeg, before quickly shifting back to showcasing the catalog's undisputed utility. "People who aren't HVAC maybe won't be excited by it, but if you are HVAC, it's your 'bible' because it takes you through all the products you need to do your jobs."

Complementing the catalog is Price's 1,400-page *Engineer's HVAC Handbook*, which explains the engineering fundamentals behind Price's products and serves as a design guide on how to execute design for heating, ventilating, and air distribution in non-residential construction spaces. To date, over 30,000 handbooks are in circulation worldwide, with the online version also available.

Feedback fuels innovation

Whether it is the PTCs, the labs, or the Price "bible," the only thing more valuable than Price's investment in innovation has been the constant feedback from its customers. Since the 1990s, Price USA has hosted its sales representatives for an annual Rep Council meeting. The gathering is a no-holds-barred event where direct and candid feedback about Price's products and services is shared directly by the sales reps to leaders at Price.

One of Price's longest-serving US reps, Kevin Harre, has served on Price's Rep Council since 1996. According to Harre, along with knowing Price's products, sales, and operations, invited council reps are also not scared to voice their opinions. In Harre's experience, the council would bring forward a list of recommendations every year. While most of the items changed year to year, one commitment never changed, and that was Gerry's resolve to complete the checklist from reps. "I always felt Gerry did a great job of listening and learning through these councils. He never took the feedback personally. I have watched Price take this information and grow their business, whether that was new products or new processes," says Harre. On a personal but worthwhile note about the human side of doing business with Price, "As hard as we all worked, we could also play hard, and I enjoyed those moments and conversations as much as the meetings," remembers Harre fondly.

More current examples of feedback from reps exist, but Gerry fondly goes back to the beginning. "It was the fall of 1991, or there about, and the Rep Council said to us that the most important thing for us to do was to change our catalog from a binder, with all the different brochures in it, to a phone-book-style catalog." That change is reflected year after year in the Price product catalog. "It's easier for the engineers to find products and engineering data on the product."

If customers have input, they will have greater interest in using the product. That innovative driver not only produces better products but also creates an enhanced customer-service experience, assures Penner. "By allowing us to advance and continuously improve our products, it really goes back to serving our customers when we're out in the market selling, educating, teaching, and training about the benefits of our products. We're continuously getting feedback, and we're always wanting to take that information back and to make our products better, so that ultimately, we can serve better."

●

Price Points on Innovation

Don't settle for good enough

Years into the development of new products at Antec Controls, the design team started over because the product wasn't as "good" as it could be. This going-back-to-the-drawing-board approach made the products better in the long run and made sales soar.

Own your supply chain

"If you own your own supply chain and become a 'one-stop shop,' not depending on suppliers with long lead times, you can ensure your customers have dependable short lead times all the time," states Gerry. Sourcing supplies to produce products can lead to a requirement of ordering large quantities with long lead times. As a company innovates, this requirement can lead to scrapping obsolete parts. During a supply-chain crisis, such as that cause by a pandemic, this issue is exacerbated. Price Electronics mitigated this supply issue by redesigning the circuit boards with substitute parts it could source, leading to no shutdowns of Price HVAC products during the time when its competitors were required to shut down due to low or absent parts inventories.

Know that winning takes time

On average, Price covers losses for start-ups for five to 15 years before seeing an ROI, underscoring an unwavering commitment to innovation. Getting the science right, learning how to build it, and marketing all takes time, but proper innovation cannot be rushed. It takes what it takes.

Train others

Once the science is sound and the products are perfected, training its customers on how to utilize Price's products is the next essential step. Price built the Price Technical Centers (PTCs) in Georgia and Arizona to provide educational and informational experiences for customers, suppliers, sales reps, engineers, and other key stakeholders.

Use feedback for innovation and to develop loyalty

When customers have input into improving products, they will have greater interest in using them. The customers' opportunity to contribute feedback not only produces better products but creates an enhanced customer-service experience. While you are out in the market selling, educating, teaching, and training about the benefits of your products, continually gather feedback to involve the customer and make your products better.

 **Risk**

GERRYISM

Being on the cutting edge means managing unfamiliar risks.

RISK TENET

Venturing into new territory brings with it adversity, frequent failures, and recovery, as you try again and again. This is the risk of building a new enterprise and striving for great heights. To get to the highs, you need to grind through the lows. You never fail until you stop trying. We are a company of grinders.

Risking US expansion, the university precursors

Price's ambitious entry into the US started to bear fruit in the spring of 1996, after seven straight years of foundation building and serious losses. Getting his long-awaited proverbial piece of blueberry pie would be imminent with the opening of Price's new plant in Suwanee, Georgia. "I was full of joy when this new plant was being built," emphasizes Gerry. At this point in the interview, the energy in the room ramps up, and Gerry sits up even taller— if that is possible for a man who stands taller than most at 6'4". The shift in body language genuinely affirms just how happy Gerry was when this plant opened. "We were about to have our first owned footprint in the States; I was just pumped." So pumped, in fact, that putting up his entire company—including the operations in Canada, his family's home in Winnipeg, and his cottage in Lake of the Woods, Ontario—as a personal guarantee to secure funding to build the plant seemed worth the risk.

Gerry's ability to detach emotionally from this tremendous risk was hard won. He took on substantial risk early in his academic career, including accepting not one but two doomed-to-fail major academic projects that, according to Gerry, were among his life's most gut-wrenching experiences, ultimately preparing him for taking unprecedented risk in business in the years ahead.

Transitioning from his undergraduate degree program into his master's program in applied meteorology, Gerry was assigned to present a research paper at the Canadian Meteorological and Oceanographic Society Conference in Winnipeg in the summer of 1970. His paper's focus was how to use modeling from computerized weather forecasting to predict weather patterns.

"I knew mathematics, fluid dynamics, computer science, and numerical simulation techniques, but I had no meteorological training—zero," recalls Gerry about this predicament. Although he knew he was not qualified to present this material to this audience on this topic, as a new master's student, he felt obliged to take on the paper. "I had no ability to say yes or no. I worked almost every single day until 4:00 a.m." As the presentation date inched closer, Gerry's research was simply not adding up. "I'm about two weeks away, and I finally came to the realization that the weather reels I worked off of were erroneous. The northern hemisphere was purported to be the southern hemisphere, and the reel for the southern hemisphere was purported to be the northern hemisphere. I was spinning up an impossible Earth. This kind of weather doesn't exist. No wonder I wasn't getting anything." But with less than two weeks until presentation day, Gerry knew he had to push through, with the task of presenting becoming more daunting with every passing hour.

On the day of his presentation, Gerry faced a room full of esteemed scientists in the fields of meteorology and oceanography. "And here I am, a brand-new master's student about to present to them on data that's flawed," laments Gerry, remembering the storm clouds forming on the horizon that he would have to weather. "And I basically knew nothing. I had to stand up there and present, which I did. Thankfully, they were kind; I will never forget that. They let me off the hook." Disheartened and discouraged, Gerry made a lifelong resolution that day that has guided him stoically through some of his most prolific expansions and advancements. "I made my presentation and exited the building. As I walked back to my car, I resolved never to commit to doing something that somebody else commits to on my behalf. Going through that kind of humiliation is game-changing in your life. You want to talk about crushing waves of heat going through your body and stepping up on stage going through that process ... honestly, that was the most humiliating thing I could have gone through in my life professionally," reveals Gerry, pausing, exhaling, and affirming almost more for himself than for anyone else, "Never again, never again. That was game-changing."

With his master's degree completed, Gerry moved with Barb to Bethlehem, Pennsylvania, to pursue his PhD at Lehigh University. While they were in Pennsylvania, their two daughters, Ainsley and Shandis, were born. Gerry would eventually complete his PhD in applied mechanics at Lehigh University, a journey that would come down to another gut-wrenching, stressful conclusion.

As described in the Introduction, Gerry's first full-time job out of university was serving as a defense scientific services officer for the Defence Research Board in Alberta, a job that required a PhD. Almost three years into his PhD program, he realized that his research would require a complete reboot. The data output from Gerry's model was not aligning with the mathematical predictions of other researchers to support his thesis's objective of determining how long it takes for a frontal system in the atmosphere to develop, or be definable in the atmosphere. Based on research from a scientist in London, England, Gerry's PhD thesis should have predicted that the weather system would develop at approximately 76 hours. Gerry's research predicted less than half that.

The timing could not have been more difficult. Gerry's National Science Foundation grant to study at Lehigh had just about run out, he was just over a month away from starting his new job that would support his young family, and Barb and the girls had returned to Winnipeg during the waning weeks of Gerry's PhD to provide him with uninterrupted time to figure out his calculations. Knowing he had not earned his PhD and was still struggling to complete it, Gerry accepted that job in Alberta, with no backup plan.

The consequences of not reconciling the calculations haunted Gerry. "No PhD means no job," discloses Gerry, recalling how he worked feverishly to bridge the gap in the missing atmospheric hours to support his thesis. He finally figured it out on a flight from Winnipeg back to Bethlehem after visiting his family. "And I thought all of a sudden about this vertical energy variable I removed, thinking it was negligible," recounts Gerry, holding his breath as if he were once again on that precipice of this PhD eureka moment. "I thought, what if I throw that back in? And miraculously, the frontal system appeared at around the required time and worked like a charm."

The risk of accepting a job in Alberta in the face of crushing adversity to complete his PhD and putting his family in potential financial harm made putting personal guarantees to secure the US expansion seem far less consequential. According to Gerry, "In all honesty, after going through what I went through in those previous years, nothing fazed me. Nothing was like what I went through in my PhD program, with my family depending on my success."

Banking on US expansion

With comfort in risk-taking established early on in his career, Gerry's decision to expand into the US seemed not only desirable, but obtainable. However, his first challenge to breaking into the US market would turn out to be one of his biggest—getting a bank loan.

"How did I get a banker to loan us money to build a plant? I talked to pretty well every bank in Atlanta. And I got 'no, no, no, no, no' because of weak financials in Price's Canadian operations and losses in the States. They wouldn't touch it with a 10-foot pole," remembers Gerry, before revealing how his persistence eventually paid off. "For whatever reason, in the spring of 1995, NationsBank agreed to the loan request. We then proceeded to buy a parcel of land in Suwanee, Georgia, and worked with a development company called Rooker on a design-build for $3 million."

With the resources secured and the plant's development underway, Gerry received a call from NationsBank in June 1995 that would have him making a big call of his own. "They said, 'Gerry, we made a mistake. Your line of credit is not approved.'" And with that news, Gerry's piece of blueberry pie was slipping away. "I had to do a personal guarantee to secure the loan," declares Gerry, remarkably calm as he recalls how his company, and his personal assets, were on the line. "I had experienced major risks before I came on board with Price. My whole life has been on the edge of a cliff. And I started to realize I'm not even aware I'm on the edge of a cliff because I'm so used to hanging on."

Banking on Barb

Imagine driving home from a meeting with a bank with a folder full of documents that itemize a loan guarantee against all your personal assets. You may be rehearsing that conversation before seeing your spouse. "Basically, everything I owned was at risk on that loan," stresses Gerry. On his ride home to see Barb, he did not give it much thought; he was weathered in risk from university and was more than comfortable with putting all his assets on the line. "Barb, being the supportive girl she has always been with everything we've done in life together, signed it without hesitation. No long discussion, no big debate," recalls Gerry.

"I just trusted him," supports Barb, and she had good reason to, citing years of witnessing Gerry prudently overseeing Price's finances. "If he did the analysis, and determined that it was the right decision, then I knew it was necessary." Remember, it was not just Gerry living through those early university trials and tribulations—so was his family.

The couple left Winnipeg late in July 1972 with all their worldly goods (other than furniture) packed into an aerodynamically built wooden box on the top of their 1970 green Maverick coupe. They planned to stay in campgrounds across Canada and the US to arrive in Bethlehem mid-summer in preparation for Gerry's PhD program and his teaching stint at Lehigh. Reaching Bethlehem on July 31, 1972, they set themselves up in a campground in the Poconos, slightly north of the city and a short drive to Lehigh, at a cost of $3 per day. This suited their budget since they had very little US cash to carry them through into September, at which time Gerry would begin to receive his National Science

Foundation scholarship and his pay for teaching undergraduate thermodynamics to engineering students.

Upon their arrival in Bethlehem, they found a one-bedroom apartment on Catasauqua Road that was not ready for occupancy but for which they entered a lease beginning September 1, 1972. The furniture from Winnipeg (sofa, mattress, kitchen table, four chairs, and an orange living room shag rug) was shipped late July and was expected in Bethlehem by mid-August. While in the campground, waiting for the apartment to be ready and for their furniture, the couple built much of their other furnishings. The furniture did not arrive mid-August as planned but appeared a month later in mid-September. As the couple was settled in the campground from August 1 on, they stayed there until their new apartment was ready for occupancy. They moved into the unfurnished apartment early in September, using an air mattress to sleep on. After a couple of weeks of camping out in the unfurnished accommodations, the furniture did arrive, and Barb and Gerry were set thanks to the furniture that they had already built.

"We had to live in a tent for a whole month without any sort of furniture whatsoever," echoes Barb, smiling warmly with the playful laughter necessary to not only survive but thrive in the circumstance. "We used the picnic table as a workbench, and we made furniture together for our apartment," recalls Barb, with Gerry by her side, eager to chime in on their early intrepidness.

"I built the coffee table with shells that Barb pulled out of the dumpster after a clambake. She set them into plaster on a plywood box, and she made the carpet underneath our dining room table with strips cut from rug samples. "I earned $400 a month from a teaching assistantship from Lehigh University and a National Science Foundation scholarship. With no family financial support, no work visa for Barb, $235 a month for rent, and two babies to support, our meals were inexpensive but creative. We ate chicken necks," adds Gerry, without even a hint of misfortune. "Back then, we would take hot dogs, cut them lengthwise in half and then into little pieces. Barb cooked the pieces in sweet-and-sour sauce and called our recipe 'Wahanee Weiner Wings,'" remembers Gerry, finding gratitude in their limited sustenance. "We were frugal, but we were happy as hell."

Given the backdrop of living lean for many of her formative years, Barb intimately understood there is no reward without risk. "I grew up in a family where we didn't have a lot of money," says Barb, expanding on how those challenges served her in supporting Gerry's US adventures. "I learned from a really early age that good taste didn't necessarily need to cost more," picking up quickly on Gerry's recollection of lifting her out of a dumpster after she had jumped in to collect clamshells she had spotted earlier at a friend's apartment during a clambake dinner. "Oh wow! I could do something with those," Barb remembers, exclaiming how excited she was with her discovery. "Gerry and I ventured over to the garbage bin after hours, and Gerry climbed into the bin," notes Barb. She is quickly interrupted by Gerry, who gently corrects her; he lifted Barb out of the bin. "That's right, you did!" confirms Barb, with a gentle jab at Gerry in a tender moment that displays their powerful bond as a married couple of 50-plus years. "Yeah, here I am saying you were more of a gentleman than you really were." Her hearty trademark laugh is soon matched by Gerry's in a cacophony of chuckles that fills the room. "Then we took them home, and we boiled them," shares Barb. "The stench was unbelievable in our apartment," remembers Barb with more laughter, shifting to the determination of the task at hand. "I embedded those shells into a frame that Gerry made, and we used that pretty cool table for the next 15 to 20 years."

The table seems a fitting symbolic output of finding opportunity in the risk of leaving the comforts of Canada for their US adventure. "We lived on a shoestring, but we didn't feel poor," offers Gerry proudly, bolstered by Barb's quick affirmation and admiration. "Good taste in design, good taste in food, and good taste in friendships. And that's what it's all about—and good taste in choosing your life partner."

Trusting in trends

With a personal guarantee signed and Barb's support confirmed, Gerry innately knew the US expansion losses would subside—eventually—with the new plant in Suwanee, Georgia. Even though others urged him to cut his losses and shut down the US start-up, he trusted in the trends he was observing. "I saw a steady ramping up in total gross profit margin, year after year, and I saw the losses growing for a number of years and then turning the corner and starting to decline," affirms Gerry. "I knew we would make it once we had achieved a year-by-year success of incremental gross profit being greater than incremental overhead. It's as simple as that. I knew the trend. I knew the data cold because I gathered the data on my own."

The bottom line, according to Gerry, was once he knew where the earnings were, he reengineered his key US products one by one and, in the process, took them from a lower margin to a higher margin. Then with the volume of sales, that higher margin grew to support single-digit growth and, eventually, double-digit growth in the teens, contributing substantial revenue into the US expansion. From that point on, Gerry and Price never looked back. "Trend is the key thing in life, really, because either you're advancing or you're not," says Gerry. "In business, where you're trying to earn off various products, you better know where your winners and losers are. We managed, product by product, taking them from being unacceptable to not only acceptable but ultimately brilliant products that were far better quality than our competitors', earning enough to keep our doors open," professes Gerry, stating that this strategy led the way for Price's successful and sustainable US expansion.

Gerry is a risk-taker, but assuredly a calculated one. "My only ground for hope was my data," evaluates Gerry, narrowing the focus. "It was the trend of the data that was important, not the absolute value. And in a start-up, you always lose for a period of time, but you don't know if you're making progress unless you see a systematic growth of gross profit dollars period by period. If you don't see that growth at a faster rate than the overhead is growing, you're done," shares Gerry, adamant in his resolution that while everyone doubted his logic, he never doubted his data and bet the company's survival, and ultimate success, on its trend toward growth. "I knew we were on track within two to three years of the start-up. I knew this was going to work. It was just a question of time. Can I last long enough? Can I keep the banks onside long enough? Can I keep the team onside long enough?"

All the while, Gerry told no one about his personal

guarantee to secure the loan. "I wasn't worried," confirms Gerry emphatically, bordering on annoyed when pressed to rationalize his lack of worry, "because the trends were so solid. The bank didn't know the trends, but I knew we were fine. I knew we had no risk. To me, it wasn't gut-wrenching to have to decide to sign a personal guarantee."

Years, even decades later, those who were closest to Gerry during those risk-infused expansion years were none the wiser. "I never knew how hard it was for him to buy that business— and how much risk he had put himself and his family at for that business," credits Kevin Harre, one of Price's longest-serving US sales reps. Harre only learned of the risk Gerry took during a casual conversation more than a decade later while visiting Gerry at his cottage. "It's an amazing story. I never knew that *that* guy who was sitting in our office saying, 'What's it going to take to get this $100,000 of business?' *really* needed that $100,000 in business."

Risk assessment

An established risk-taker, Gerry's decisions were not made without the meticulous calculations of a man steeped in mathematical erudition. As confident as Gerry was with his US expansion and with considerable investments in the present day, he will tell you that if the trends and data do not support the investment, it is risky business not worth venturing into. And had his trends not supported the US expansion, it is unlikely you would be reading *I'm Just Gerry* today.

Joe Cyr, who has been along for some turbulent adventures, riding the wave of risk with Gerry over the years, is keen to point out that Price will take risks, but with prudent assessment. "It's more what I would call managed risk. I think sometimes when people talk about risk, they think of it as wild forays into the unknown. In 90% of our cases, we're going into adjacent territory."

Respecting the blazing trail to the US Gerry scorched in the 1980s and 1990s, Price has learned from the past and commits to protecting the future. "We never take what you would call balance-sheet risks," attests Cyr. "We don't put the company at risk. It would be true, certainly in the early '90s, that Gerry was awfully close. But remember, he's a modeler by background, and he was always modeling the business financially. He knew where we were in terms of risk profile and what we needed to do to get there." With that track record of prudent risk assessment engrained into the history and culture of Price, the company will continue to pursue growth by taking risks, albeit not likely scenarios that require any more personal guarantees. "They're not bet-the-farm risks. And they're not balance-sheet risks," affirms Cyr.

●

Price Points on Risk

Risk more when it is required

After months of rejections from many financial institutions, Gerry finally secured a loan from a US bank to purchase a plant in Suwanee, Georgia, only to be told shortly afterward by the same bank that his line of credit was not approved after all. Calm and assured, Gerry offered his company and his personal assets as a personal guarantee to secure the loan. He was willing to take a calculated risk.

Trust in trends

Gerry trusted the trends he modeled. Seeing a steady ramp-up in total gross profit margin year after year, he still projected losses growing for a number of years—but not indefinitely. And as he predicted, Price turned a corner with losses subsiding and with incremental gross profit being greater than incremental overhead.

Understand risk assessment

As a confident risk-taker, Gerry carefully assessed risk: if the trends and data do not support the investment, it is risky business not worth venturing into.

# Resilience and Grit

GERRYISM

We are a company of people who grind their way through difficulty, setbacks, and adversity.

RESILIENCE AND GRIT TENET

Like the stonecutter who chips away at a rock, not knowing when it will break in two, we persevere at our tasks, not knowing if our effort today will succeed or not, staying on task until we eventually prevail.

Under attack

Valentine's Day, February 14, 2019, 6:29 a.m. The scene inside Price headquarters in Winnipeg was anything but romantic. It was complete chaos. IT staffers frantically pulled cables from computer servers. Every unplugged computer and disconnected cable lessened the chance of valuable data and systems becoming corrupted. Price was under attack. Cyberattack.

Despite the efforts of every able set of hands, Price began to shut down. Seconds turned to minutes, and minutes evaporated into a few desperate hours. Eventually, every system in every plant and facility came to a complete and utter standstill. Twenty-seven facilities. All operations down.

"We're under attack!" reads Cyr. The text message is saved on his phone—and seared into his memory. *"We are shutting everything down."* Five minutes after getting the text, operations went dark. Cyr remembers getting the chilling confirmation of the deliberate attack embedded in cryptic ransom messages in Price's server systems.

"This was scary beyond words," recalls Cyr, reliving the moment, evidenced by his breaths getting shorter and details spilling out as fast as he can share them—as if he were standing right there in Price's headquarters reading the message for the first time. "When you read that cryptic message in broken English that basically says, 'We have your systems, and if you wish to have it restored, pay up.' They were asking for ransom. And, I mean, that just seemed unconscionable to us." For US$2.5 million, paid in Bitcoin, Price's systems would be freed. "It's a tremendous amount of money. We're not going to do this. You know, we've got to try to stop them."

February 14, 2019, 8:00 a.m.

Cyr sat at a boardroom table, flanked by a hastily assembled team of IT security experts. In jeopardy were Price's major digital libraries that represent thousands of products and millions upon millions of pages of data, including intellectual property (IP) and patents. "My main concern was our IP that relates to product testing," divulges Gerry. "That's all the lab data, and that performance data goes back to 1978. We had 40 years of engineering records in that library."

With facilities dark and production halted, the first step was to "know your current reality; know it now." This Price mantra was routinely spoken at operational assessments, yet never was it more important to call upon than it was at that moment. "Because the plugs were pulled quickly enough, the destruction didn't get past the letter B of our library alphabet system of files," explains Gerry, with a big exhale. He feels as grateful today with that understanding as the moment he first heard that assessment. "It takes processing time to destroy it all."

Concurrently with the data damage assessment, Price informed its key stakeholders—customers, employees, suppliers, and more. "There was no 'Hey, we've got a bit of a problem here' or mincing of words," asserts Cyr. "I had to jump on a plane to go to Atlanta immediately. I will never forget showing up in our big factory with the lights off and all the systems down. It was too quiet."

With a full shutdown in place, Gerry leveraged the internal expertise around him, complemented with a cadre of consultants from all over the world, especially an expert that translated the continual cryptic conversation with unknown and clandestine cyberterrorists holding his company hostage. "We negotiated right out of the gate. We had a plan A and B," outlines Gerry. Plan A: restart on our own, figure it out, and not pay the ransom—if for no other reason but to buy valuable time on Price's end to truly assess the damage and calculate how long it would take to be operational again. And plan B: if plan A failed, entertain all other options because leaving customers hanging indefinitely was not an option, even in the short term.

Gerry's prime concerns were the following: "What could we recover from backups to restart the business? And how quickly could we restart our HVAC plants?"

February 18, 2019, 2:00 a.m.

Regardless of the strength of the attack, with round-the-clock meetings and countless trips back and forth to Price facilities in Atlanta, Phoenix, and Calgary, the resilience of the organization shone through. In the early hours of Monday, February 18, 2019, the company was operational—albeit with significantly reduced capacity. Nonetheless, running from backups of data and reduced in capacity as a result, the Winnipeg plant was the first to hobble back into operation. Atlanta's Suwanee plant followed in the same hampered fashion the following week, and at only 20% of its regular capacity.

The Georgia facility was heavily reliant on automation and so was rendered practically unusable with the computer networks down—down but not out. "We moved to full manual," recollects Gerry. "Everything, including labeling. We had to figure out a way to bypass our whole system but still use enough automation to make it look professional." In Atlanta, a gritty effort where everyone, from C-level officers to frontline staff, was pitching in, printing and manually adhering labels to products and packages. All the while, Gerry, Cyr, Fraley, and their teams continued to negotiate with the cyberterrorists to buy time, not knowing what the attackers' next move might entail. "We countered a couple of times to maintain the appearance of the dialogue," offers Gerry. "We still weren't sure whether we would be paying the ransom. It all depended on our due diligence and what we couldn't restart."

To this day, the exact location of the cyberattackers remains unknown, a factor that made the days and weeks that followed the attack all the more stressful. "I was negotiating with them through an intermediary, and it was just scary," expresses Cyr. As days passed, tense negotiations continued as well as the uncertainty. "They reduced their demand from $2.5 million, expressed in Bitcoin, down to $2 million. We were trying to play the game and keep them on the line and to not provoke them into doing anything worse."

More importantly, the time tactic allowed Price to assess damage, reinstall backups, and build stronger cybersecurity at the same time. "Literally the day of the cyberattack, we began a complete changeover to the advanced computer network system that's in place today," avows Cyr.

While the cyberattack did not penetrate as deep within the company's digital libraries as feared, the damage to security and operational systems was still extensive, requiring an ascension of resilience and grit not experienced by Price since the heady US-expansion days. "Over the course of the two months, we met every single day," reports Cyr. "We would meet with our plant managers, our executive team, and each affiliate company. It was like a battle." And with systems slowly restored to full capacity and cybersecurity bolstered, Price's cyberattack came to a momentous end, eight weeks after it started. With the engineering team's confirmation of full restoration, Gerry never paid the cyberattackers a dime or, in this case, a Bitcoin. "We dragged negotiations on just to keep it alive as we have a very complex system of hundreds and hundreds of major servers," justifies Gerry of his cyberwar victory. He wanted to provide as much time as possible for his IT and engineering teams to fortify cybersecurity. "Our IT teams then took us to a completely different level of cybersecurity in all our facilities across North America."

Crises can bring out the best or worst of an organization, and while Gerry would not wish a cyberattack on any company, the response to Price's cyberattack remains one of his proudest moments, most notably the resilience and grit of everyone who came together, not relenting until the threat had passed.

Build a boat, make a plan

If there is a top-10 list of the most stressful experiences leaders and their companies could survive, a nefarious cyberattack would most certainly be one of them, which Gerry will attest to from his perspective of a 37-year career leading Price. "What I do when tough situations happen—I make a plan," qualifies Gerry. "I don't cave. Generally, it takes a couple of days, never more. It's just a conceptual plan. Once it is in place, or roughed out, I feel good." As important as getting an action plan in place is to Gerry, much like the cyberattack plan, the shift in mood gives Gerry and his leaders the energy to focus on what is Important. "Now, I'm not saying the execution is simple, but I'm determined. And I deliver on every single element of that plan with laser focus."

That focus, stemming from his growing-up years and his time in university, has served Gerry faithfully. "I heard it from my mom and dad," remembers Gerry. "The concept that you're a completer and that you finish things was ingrained in me." Recalling the basement boat project, "Once I started that boat project, I wasn't capable of scrapping it halfway. I had to see it through. I'm downstairs building this damn boat. I just have to finish it. If you started it, you finish it. I think I caught on to that early in life. You don't quit halfway just because it's distasteful, taking up too much time, or taking you away from other things you want to do." In addition to serving as a metaphor for resilience and grit, influencing all operations of Price and its affiliates, building that boat paid off romantically for Gerry as well—sailing away with his teenage sweetheart, and love of his life, Barb.

Grinding out success

"The one word I always use to describe Gerry is relentless," proclaims Maykut, affirming Gerry's resilience and grit have a cascading influence over all businesses and operations. "There's just a mindset that we will never stop. Maybe there is one of those 10 projects you should have given up on. However, the nine out of the 10 successful outcomes that get you to the finish line probably make up for the one that doesn't," estimates Maykut, "Just like Gerry, just never give up. Always look for a solution and just be relentless."

Recalling an American Society of Heating, Refrigerating and Air-Conditioning Engineers (ASHRAE) conference in Chicago where Price was hosting dozens of reps for social activities, Fraley recalls how the festivities went into the early hours, and there's Gerry, going over a file with a few reps in an adjoining room at one o'clock in the morning. "He was always this track star you could never catch," recalls Fraley, shaking his head with amused disbelief. "He was effusive. He was always on it."

The resilience of Gerry's determination in the US-expansion years has become one of Price's tales of folklore. "He was so hands on with the Atlanta operation," praises Cyr about the nascent stages of the US expansion. "He would fly to Atlanta on a Sunday or Monday and come back on Thursday, week after week after week, flying overnights or on evening flights because he needed the daytime to be working with operations and sales teams and, of course, visiting reps." From Cyr's recollection, Gerry's red-eye adventures, so to speak, lasted for many years until Atlanta's operation started to become profitable. "He has an uncommon stamina, uncommon resilience, and staying power. It's become a part of our company's fabric."

The long view takes resilience

Many leaders featured in *I'm Just Gerry* profess one of the leading attributes of success unique to Gerry is his determination. "It's amazing where the company is going right now," evaluates Gerry, during his April 2022 Bankers' Orientation in Winnipeg—a fitting time and place to extol the virtues of resilience and grit over the long haul. "Using robot arms, we've automated many production functions normally done by people. If it's essential, we automate," adds Gerry, elaborating on how it is vital to remain committed to the process of automation. "The reason I'm pointing this out is because its common for us to try something out to be more automated, and we kind of die on the sword because we can't quite figure it out, but over six or seven years, we do. And it really paid off this past year when demand doubled. We buy equipment ahead of demand is what I'm trying to say here, and sometimes it takes us time to incorporate it into our business operations."

The manufacturing facility in Price's plant in Winder, Georgia, is just another example of seeing the long view, espouses Fraley. In this facility, Price manufactures air moving products that include terminal units and fan coils. "We opened the Winder plant, located in Winder, Georgia, in the fall of 2008, right as the USA entered the subprime-market meltdown and Great Recession. The market was crashing, and people probably thought we were crazy. Without that plant expansion plan, we would have failed to serve our customers in the US, but nobody knew our intentions to expand at the time," remembers Fraley, remarking on how resilience and grit got Price to this level of success. "Gerry invested heavily in the noise control lab, and it would have been easy to give up. It would have been easy to say we misunderstood this, or we miscalculated, and let's just get out. But we don't quit."

Another example of Price's resilience and grit is its entry into the specialized rooftop air handler business. A rooftop air handler works to distribute air to specific areas within the building it is servicing through connected ductwork. "We entered the rooftop air handler business in 2010 with SolutionAir," describes Cyr. "More than a dozen years later, we're into this now for several millions of dollars. We've sure made a lot of mistakes, and we're still finding our way during these early years of product development. You may wonder, with things going well in other parts of the business, why we decided to continue to give it our best shot, to not back away? Not Gerry, not Marty, not us." Cyr adds, "The last piece of that puzzle was finding the right leader for SolutionAir to straighten out the operation and put us on the right path, which we've now done."

Based on past performance of long-term commitment and long-term investment, Cyr anticipates the rooftop air handler investment will generate a significant return on investment. "It's easily the most heavily invested part of Price." As the company explores new territories in new markets, its leaders anticipate pushing past challenges and leveraging the opportunities in years ahead. As Gerry likes to say, "There are no shortcuts."

●

Price Points on Resilience and Grit

Use crisis as opportunity

A cyberattack shut down Price's operations across the continent, putting the Price Way's *Resilience and Grit* Tenet to an extreme test. The team assessed the true damage, rebuilt, and used the overall opportunity to evaluate and strengthen security systems, mitigating future attacks.

Make a plan when the going gets tough

When faced with villainous challenges such as a cyberattack, instead of getting emotional or frustrated, focus the stress into making your plan. Execution may not be simple, but without a plan, success will be less likely.

Pursue success relentlessly
Whether it is surviving a cyberattack or researching and developing products with intricate new technology, remain relentless in the pursuit.

Take the long view
One of the leading attributes of success unique to Gerry is his determination to succeed, investing years of development and millions of dollars. Taking the long view and pushing through difficult challenges are vital to building a forever company.

 # Foundation and Legacy

GERRYISM

We are striving to build a forever company.

FOUNDATION AND LEGACY TENET

We are builders. We are striving to build a "forever" company,
operated in a principled way and valued by our employees,
customers, and communities. We take the long view to success.
A successful year is one we can look back on and be excited by
the foundation we laid for the future. We define foundation as
anything that helps us serve our customers better.

Building for the future, forever

In the early 1990s, Price was reeling from the onset of a brutal Canadian recession that had plunged the company deeper and deeper into debt across all its operations. Desperately trying to stay financially afloat amid soaring interest rates, manufacturing companies were particularly hit hard, and Price was certainly no exception. Instead of reducing debt and pulling back expansion as business leaders around him scrambled to do, Gerry decided *now* was the best time to go to the bank and ask for $2 million more. "I know we're at the top of our line of credit," Gerry remembers telling his influential and accommodating banker during what would eventually become a company-defining moment. "I said, 'But this is different. This is for Alcan-Price.'"

Let's walk through Gerry's rationale for asking for the loan of millions of dollars when Price was already millions of dollars in debt.

In the early 1970s, Price did not have aluminum-extrusion science in-house as it was a sophisticated discipline that was also expensive to develop. In 1961, Gerry's father, Ernie, acquired a Titus license for Canada. Titus was a major US HVAC manufacturer at the time. The purchase came with an old aluminum-extrusion machine from Titus's factory in Waterloo, Iowa, and many of Price's HVAC products needed the extrusions to facilitate their fabrication.

Short uniform cylinders of aluminum alloy, known as billets, are pushed through a furnace at temperatures just under 1,000 degrees Fahrenheit. Once heated, the billets go through an extrusion press with 10 million pounds of hydraulic pressure, exiting as a specific profile or shape. Look at your window, and you will see an aluminum extrusion securing the glass. Look inside your car, and you will lose count of the intricate aluminum shapes holding essential pieces of your vehicle together. Producing extrusions creates almost infinite possibilities for manufacturers across many industries, including HVAC, automotive, construction, and more.

Intuitively, an extrusion process was foundational for Price's growth and service, so much so that in the early 1970s, Ernie and Gerry Law formed a long-standing partnership with the Aluminum Company of Canada, more commonly known as Alcan. The partnership's primary objective was to produce and sell aluminum extrusions in western Canada. Price contributed its extrusion press in Winnipeg, know-how, and sales acumen, whereas Alcan contributed the science of extrusion and a new extrusion plant in Calgary, Alberta, which also included an extrusion press. That 50/50 partnership, known as Alcan-Price, remained successfully in place for almost 20 years, until Gerry jumped at the opportunity to become an outright owner at 100%.

With the recession taking its toll, Alcan decided to divest its downstream operations in Canada. As the recession wore on, Price's Canadian operations showed signs of stabilizing, even though US operations were still losing money. "We still had huge debts. And our debt-to-equity ratio was enormous," recounts Gerry. He recognized the mounting debt but was encouraged by what he saw as the positive trends. "We were at the top of our line of credit, but because the trends were good, we were in better shape than we looked on paper." A trip to Alcan's headquarters in Montreal, Quebec, yielded a one-page deal where Price would assume all risks in return for Alcan selling its 50% stake in Alcan-Price. A second trip to Montreal a short time later secured the deal.

With the deal in principle, Gerry presented the bank with a detailed plan to pay back the $2 million loan within five years, with the first half to be repaid within the first two months by leveraging a line of credit that was currently in place in Alcan-Price. The plan was approved by the bank, and Gerry received the loan within a week of making the request.

"We set the closing date and had a closing ceremony with Alcan," remembers Gerry, "And we signed, sealed, and delivered on June 1, 1994. It was bought at a time when we actually shouldn't have been spending a cent on it because we were fighting for our lives, losing money in the States—but a window of opportunity was open, and it was worth acting on." In 1994, Alcan-Price was rebranded to the name it is known by today—APEL Extrusions Ltd., or APEL. The return on investment of a $2 million loan is now only a distant but substantially noteworthy memory, with APEL now valued into the hundreds of millions of dollars, a foundational example of how Price is striving to become that forever company.

"We grew Calgary, and we expanded," says Mike Flynn, APEL President, speaking of some of the big-ticket growth items from the past 20 years alone. "We bought a US facility in Oregon, and now we're building in Phoenix the largest and most technically advanced aluminum-extrusion and finishing facility in North America."

This factory is now up and running and shipping product. Whether it is the Tenet of *Foundation and Legacy* or the other Price Way Tenets, Flynn subscribes the business value of APEL adopting the Price Way. "It's good business and good values, dealing with people in a transparent way," insists Flynn, feeling confident Gerry's legacy will serve APEL and Price for the next 75 years and beyond.

Foundation, not profit, is the bottom line

Spend a few minutes exploring any internet search engine and you will instantly discover a bounty of sage advice about the value of putting culture and service before profits. Many of the leaders interviewed in *I'm Just Gerry* who have notable executive leadership experience prior to joining Price point out that embracing leadership philosophies, or in Price's case, the Tenets, is the exception and not the rule within corporate industry.

Gerry *is* the rule when it comes to building business foundation according to these leaders. "I used to be in mergers and acquisitions. One of my titles was Director of Business Integration," says Flynn. He notes that it is never easy to integrate existing companies into a new company through a merger acquisition. Flynn notes the Price Way has become a natural and intuitive way of doing business. "I would argue that just about any Price manager could come into APEL and the culture wouldn't be a big shift. And I could go into just about any Price business and it wouldn't be different than how APEL rolls."

Build the foundation and the legacy will follow, advises Gerry. "We have built the foundation of our company, our investments in people, technology, products, and plants. I don't really focus on the bottom line. I focus on what we can do that will drive future growth for future success."

No service without foundation

AROW Global, an affiliate Price company, has become a leader in vehicle window and driver-protection systems. In 2021, the company expanded, building a tempered-glass facility at its plant in Mosinee, Wisconsin, with the objective to simply serve its customers faster. Gerry believes that without building foundation, you cannot serve your customers, and any investment in foundation will pay off in the long run. Having calculated the ROI on the AROW Global expansion, no immediate return was found. "There's no return on investment there," confesses Gerry, then immediately drove his point home. "My point is, if you look at all the things we do, we invest to serve better. That's all we do. Building foundation means doing whatever it takes to serve better, whether it be new plants, automating businesses, vertically integrating—which is why we refer to ROS as return on service, not return on sales. That's what matters to us; return on investment, not so much. Return on investment comes along for the ride when you focus on out-serving the competition."

To support ROS requires Gerry and the Anvil team to trust their leaders and their teams to build foundation, even before a track record of sales has been established. That belief system is motivating to leaders such as Dan Koschik, President of AROW Global. "Leadership allowed us to invest all that money before we knew what the return was going to be. They bought into our vision and trusted us to be reliable decision-makers."

The foundational power of knowledge

A fundamental component of the *Foundation and Legacy* Tenet is knowledge gained through trial and error, especially when entering new markets and developing new products. "When we enter a greenfield, there's no point in doing it to become small; you do it to create a foundation so you can scale it to the nines and go up to the stratosphere," says Gerry.

The challenge for most businesses attempting to build foundation is that it does not always produce ROI because the strategy of generating sales obfuscates the value of producing foundation, observes Cyr. "That's foundation. You can't see it, you can't touch it, and it's not on your balance sheet. Foundation is a lot more than the things that most businesspeople would define as knowledge; it's know-how, it's experience, it's grit, and it's learning from mistakes." On the path to success, there will only be failure if there is no knowledge gained. Cyr adds, "A successful year is not always based on the income statement. It's based on how much foundation you build because that foundation will generate the financial results you need to grow tomorrow."

The Price Paradox

At the end of the day, *I'm Just Gerry* is ultimately a book designed to educate and serve internal and external stakeholders. By that measure, it is also an exercise in the Greater Good Tenet, sharing foundational knowledge for the benefit of all, regardless of whether readers become customers, employees, suppliers, or partners.

I'm Just Gerry also honors Price's 75 years in business (1949–

2024) and Gerry's 75th birthday. Preserving Gerry's legacy through building foundation has become increasingly important for Price. Essentially, Gerry and his leadership team want to ensure the company's stories of failure and success, documented in *I'm Just Gerry*, serve generations of stakeholders in the years ahead, and to manifest building that "forever company."

Do the company leaders of tomorrow get it? "Yes, they do!" emphasizes Cyr. "There's no question that what differentiates us is the very fact that we will build foundation. We will build it right. We will build it from the ground up. The need to build foundation is absolutely internalized and part of the fabric of Price."

The Price Paradox is that as the company grows exponentially, largely under its once-in-a-lifetime leader, Gerry, it concurrently is striving to sustain that growth for years and decades into the future led by the guiding 13 Tenets that make up the Price Way. Maykut, Gerry's successor, has no plans to "replace Gerry" but rather wants to ensure Gerry's legacy and distinctive business philosophy are documented and clearly understood—now and forever. "It's not about Gerry or about Price. It's about serving our customers. If they have success, we will have success," elaborates Maykut.

Price Points on Foundation and Legacy

Build with an eye on the future
Even as a brutal Canadian recession plunged Price into record debt, Gerry negotiated a $2 million loan to buy out Price's partner in aluminum extrusion. The current financial situation should not deter good long-term financial decisions.

Focus on future growth, not the bottom line
Build the foundation and the legacy will follow. Price has built the foundation of its company, along with its investments in people, technology, products, and plants with a foundation-first focus.

Embrace learning over sales
The challenge for most businesses attempting to build foundation is that they do not always produce ROI, even in the long run. A sales focus obfuscates the value of producing foundation. Success in building foundation is measured not in sales but in knowledge gained.

Recognize the Price Paradox
Even as the company grows, it must learn for the future to avoid mistakes of the past. The knowledge gained through invaluable lessons learned and best practices should be codified for the benefit of key stakeholders.

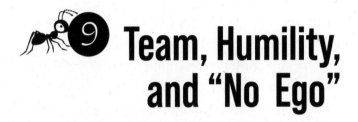

Team, Humility, and "No Ego"

GERRYISM

All our accomplishments are team accomplishments.

TEAM, HUMILITY, AND "NO EGO" TENET

We never keep score of how often we go the extra mile for our customers. Price has a "no ego" culture where we each contribute our best individually to the team win, celebrating new ideas and successes wherever they come from. Though we are high achievers by nature, we maintain our humility by remembering our roots as a small company.

Wait for the applause

For those who are born and raised in Winnipeg, Manitoba, embracing the prairie life experience earns you a badge of honor among kindred spirits—possessing a special kind of resilience and grit, tacitly understanding and appreciating the character of all others who grew up in one of the coldest winter cities in the world. If you are a proud Winnipegger, "born and raised" becomes the unofficial descriptor for those who have lived it and loved it, just like Price's Dan Koschik. "Winnipeg boy, born and raised," offers Koschik enthusiastically, "I went to the University of Manitoba, earned my engineering degree, followed by an MBA." Koschik is as devout a Winnipegger as they come, giving you a greater appreciation for the courage it took for him to then move his family 650 miles to the United States in 2006. "I picked up my wife, who was also a Winnipegger, and our two kids, both born in Winnipeg, and we moved to Wausau, Wisconsin."

AROW Global is in Mosinee, Wisconsin, a city that is a part of the larger Wausau metropolitan area. Since 1965, the company has designed and manufactured windows and glazing systems for the mass-transit industry in North America, expanding over the decades to serve operator cabs in the heavy-machinery industry and service vehicles such as delivery vans.

Similar to the Alcan-Price partnership described in the chapter on Foundation and Legacy, AROW Global grew out of a long-term, 50/50 partnership, this time with the Winnipeg-based Koberstein family's Stormtite window business. In 2005, Price acquired the Koberstein's 50% interest, and the company name was changed to AROW Global.

Koschik had just been promoted into the role of general manager at AROW Global, requiring a move to Wisconsin to oversee operations. But Koschik wasn't promoted so much as he was thrust into the role. According to Cyr, "We had no one else who was qualified to take over AROW Global after a previous leader led us astray and a key customer went bankrupt."

Cyr's admission in no way diminishes Koschik's skills or willingness to step into a challenging role, one that quickly revealed itself as more demanding than originally thought. "It soon became apparent that I wasn't prepared to handle the circumstance," recalls Koschik, not mincing words when it comes to assessing the situation. "The circumstance was way more of a shit show than I think anybody thought." The frank recollection by Koschik is perhaps an unfair self-assessment, but it is his nonetheless. "Honestly, things operationally went from bad to worse over the first couple of years that I was the General Manager." According to Koschik, and respectfully confirmed by Cyr, AROW Global's directive was to integrate a new category of business into the company's operations. "Exacerbating the circumstance that really led to the shit show was attempting to diversify from our core business of making windows," remembers Koschik, noticeably becoming introspective when recalling this challenge.

AROW Global had won a considerable volume of new bus-component metal-fabrication business and was faced with the challenge of stabilizing an existing business under new leadership. At the same time, AROW Global's team was incorporating competencies far removed from the window business that had matured over the years. "We had no infrastructure and no experience doing sheet-metal fabrication and welding," admits Koschik. The issue was not a lack of support from AROW Global in Winnipeg under Price—that was certainly present. The issue rested with himself. "I didn't appreciate the complexity associated with managing AROW Global's operation, and I didn't understand the leadership competencies that I needed. It was terrible," says Koschik.

As the situation became operationally untenable at AROW Global in Wisconsin, Cyr made the difficult decision to have Koschik take a step back while an experienced business-restructuring consultant, or "hired gun," was brought in to stabilize the business and to provide mentorship and support to Koschik and his team.

With humility and setting his ego aside for the good of the organization, Koschik embraced a new role as Vice President of Sales. As AROW Global stabilized, building foundation and increasing sales, Koschik gained valuable experience to understand the business and, in the process, became a valuable leader for the future.

Eventually, the hired gun packed in his pistol and moved on, with Koschik ready to take another shot of his own. "I remember Joe saying that the AROW Global team would very much be in support of that notion," shares Koschik, pausing to reflect on his gratitude in becoming the AROW Global General Manager once again. "I was very proud of that."

Within a few years, under Koschik's renewed leadership as GM, the company was stable and had grown four times from its original value. Not long after, Gerry brought together the AROW Global team for a business dinner in Winnipeg, where he not only recognized Koschik's commitment to team, humility, and "no ego" but rewarded him for his character and competency, announcing that Koschik would become the new President of AROW Global. The announcement was a surprise to everyone except for Cyr but was quickly embraced by everyone. "He made me the President of the company. My whole team was there. And it was honestly wonderful," remembers Koschik.

The promotion announcement was made even more special by the spontaneous and thunderous applause from Koschik's entire team. "I remember the applause," says Cyr. "I was so moved because first of all, he stepped up into a role when we had no one else. He realized he wasn't ready for it and willingly and happily stepped back. And did he ever shine in that vice president role before earning his presidency!" boasts Cyr with admiration. "Like all the things that go with humility and 'no ego,' following the Price Way, Dan Koschik is just a perfect story."

A textbook case study in humility

When it comes to humility, Gerry is literally and figuratively a textbook case, and one leaders like Koschik say they try to emulate. As you read in the chapter on Innovation, Gerry was instrumental in creating the Price catalog and *Engineer's HVAC Handbook*, both of which are heavily referenced by engineers around the world. Gerry's academic achievements include a master's degree in mechanical engineering and a PhD in applied mechanics, yet when it came time to include his credentials in the publications, he bristled at the idea. "He is credentialed, awarded, revered, and admired, universally," lists Greg Loeppky, Price's

Vice President of Marketing. "Still, to him and others who know him, he's 'just Gerry.'"

Gerry's understated and humble approach became a bit problematic for Price when first publishing the *Engineer's HVAC Handbook*. "Gerry was involved at every stage of the 1,400-page, technical textbook, cover to cover, finding and highlighting some errors, which we corrected," recalls Loeppky, contemplating just how impactful Gerry's contributions to the engineering profession have become. "It's a true gift from Price—and Gerry personally—to the industry." The problem was when Gerry was writing the textbook's foreword, Gerry could not be convinced to include his academic credentials over his signature. "He wanted to sign it as he always does—just Gerry, no credentials," notes Loeppky. "I argued with Gerry continually, trying to help him understand that as academic work—literally a textbook—he had to; it was the right thing to do. Truthfully, I'll never know if he ever agreed; I don't think he did. But we did it anyway, and he didn't complain because it was a team decision, even though he didn't agree with it."

What's in a name?

Gerry's humility and "no ego" regarding his academic credentials did not end with Price's textbook. In the fall of 2019, Gerry's alma mater, the University of Manitoba, renamed its Faculty of Engineering to the Price Faculty of Engineering, but it took some convincing within Price to get Gerry to agree to sign off.

That fall, the University of Manitoba had approached Gerry and Barb with a bold proposal to create a $20 million endowment.

The endowment's key objectives include recruiting and retaining several professors in perpetuity and creating strategies to increase enrollment of undergraduates by 25%, with a particular focus on increasing female enrollment.

Through the Price Foundation, Gerry and Barb donated $20 million to fully fund the University of Manitoba's ambitious target. Not anticipating such an extraordinary act of generosity, the university responded by requesting to rename its engineering faculty in Gerry's honor, not only for his donation but for his immeasurable contribution to the engineering profession.

An ocean away in Italy at the time, Cyr remembers excusing himself from a dinner to take a call from Gerry to discuss the university's renaming overture. "Gerry was really and truly perplexed and surprised," remembers Cyr, providing additional context to Gerry's genuine bewilderment. "It had never dawned on him that they might want to do this. They had just told him that they were so overwhelmed with his and Barb's generosity and game-changing gift that they wanted to name the faculty in his honor. He was calling me literally to ask if this was a good idea because he was concerned that it might be misinterpreted as something he was seeking—which he clearly wasn't." Cyr worked closely with Gerry for over two decades and was privy to a number of substantial donations from Barb and Gerry, many of which have never been made public. "He doesn't have an ounce of ego when it comes to giving, never looking for his name in lights or to be recognized. Knowing this, it took me less than a second to confirm that this was in fact a very good idea and good for all parties—the community, the university, the province, and Price—even though the latter was never an objective," states Cyr, adding Gerry's lack of ego never ceases to amaze him. "I found it quite amusing—typical Gerry."

Not that he needs it or goes looking for it, but if Gerry ever wanted further affirmation about the benefits of having the faculty renamed in his honor, he would only need to step out of his office and walk the hundreds of thousands of square feet of Price's facilities, especially in Winnipeg. "In Winnipeg, it's almost the pinnacle of an engineering career to be able to come to work for Price," notes Graham Fediuk, Price's Director of Engineering Services, who is one of the dozens of University of Manitoba engineering graduates who work for Price. "To have Gerry's name on it, that's just a huge sense of pride and gives me more of a sense of responsibility as an engineer," beams Fediuk. "I feel like I represent engineering across Winnipeg now because of Gerry's direct involvement with the engineering faculty, and many engineers at Price feel the same way."

Trust is built on humility

As evidenced by Fediuk's comments, a good leader's demonstration of humility and "no ego" is consistent and authentic. That is why for leaders like Maykut, effective leadership starts with humility. "Good managers and good leaders are always aware of others, and it's always just a good reminder that we don't need to be perfect." If egos were not kept in check, Fraley contends, it would discourage US reps from offering valuable feedback, thereby changing the trajectory of US growth from upward to downward. "They are completely candid and forthright with their feedback—good, bad, or otherwise—about how Price is serving them as reps."

"We default to trust," stresses Gerry, yet acknowledging many situations where trust was not reciprocated. "We never default to suspicion and will always believe until we have evidence that we shouldn't. I refuse to be suspicious."

No egos on the front line

When gauging the sentiment of Price's culture of humility, team, and "no ego," Gerry looks no further than the front line of his operations and has consistently mentored his leaders to do the same. There you will find, says Gerry, irrefutable displays of genuine humility. "The ones that I have the most respect for are our frontline workers in the plants. They show up and willingly and relatively joyfully slug it out. On the production line, they're there and they can be counted on to deliver the goods." Gerry pauses to ensure he is effusive in his praise of their contributions. "They put up hard work every day, and they keep the wheels on the bus for the company. I love these people. To me, they are the salt of the earth, and they deserve our greatest respect and admiration."

Frontline worker Bill Cross, a retired Price Industries factory worker who worked a variety of positions over his 40-year career, says the respect for Gerry is mutual and that Gerry earned it the hard way right from the beginning. "Being the owner's son, we tended to give him some of the not-so-great jobs. But you know, he never complained, and he learned the whole experience of the shop floor. I think he understands from a very direct perspective exactly what it was like to work on the shop floor." As Gerry climbed the corporate ladder, he was always available, notes Cross with appreciation. "Whether you were a spot welder or you worked in the office, it didn't matter—Gerry was always accessible, regularly seen on the factory floor throughout my career."

Accordingly, Price invests in continuous-improvement measures to elicit insightful feedback and to encourage innovative ideas. "We really empower the frontline workers to tell us the issue, dig in a little bit on their own, find the root cause, and then propose a solution," shares Jodi Tomczak, Price's Vice President of Human Resources. Tomczak confirms there were upward of 2,000 ideas submitted a year, just under 18,000 ideas in total since 2007, all thanks to Price's frontline staff. "It's a testament to the entrepreneurial culture of Price and how Gerry empowers people all over the organization."

For Gerry, empowering his frontline teams means genuinely implementing their ideas. "We respect our frontline people, cultivating from them ideas to make their jobs easier and to take away the strain and pain of what they do. They have so much wisdom because they do it every day. We collaborate with them and leverage their wisdom to make changes to keep our company sharp."

Humility, for now and forever

This chapter is a demonstrable representation of how the company is ever aware of its humble beginnings as its touchstone, anchored in its fervent belief of the Price Way's Tenet of *Team, Humility, and "No Ego."* Today, the company is experiencing unprecedented success, and while the culture recognizes achievement, its leaders proclaim that humility may just determine if it becomes a "forever company."

●

Price Points on Team, Humility, and "No Ego"

Know when to step back

A leader was asked to step up into a role but soon saw that he did not have the competency to complete the goals of the organization. Putting the company's success above his own, he stepped back, allowing others to complete the process.

Build trust with humility

Effective leaders start with humility, releasing any sense of ego in an effort to build trust with their teams. It is this releasing of ego that allows leaders to listen to everyone in the organization—frontline staff, customers, and other leaders—to grow the organization.

Empower the front line

Empowering your frontline teams means genuinely implementing their ideas. They know best how to improve their work situations. They are working with the products every day and may have input that no one else has.

Embrace humility, now and forever

Price's humble beginning is largely anchored in its Tenet of *Team, Humility, and "No Ego."* Humility may just determine if an organization becomes a "forever company," because if you think you are above improvement, you soon become arrogant and complacent.

 # Execution

GERRYISM

We work in real time, fulfilling urgent customer needs nearly "instantly," which requires us to be "current" at worst, "ahead" at best.

EXECUTION TENET

We have a high-speed execution model at Price. It is rooted in quick action, course correction, and knowing our key drivers of success. By pursuing a small number of game-changing objectives each year, we can easily track our performance, counter roadblocks, and respond to customer needs.

A Price pandemic pivot

In early March 2020, the COVID-19 pandemic almost suspended the North American economy. Many industries ground to a halt, forcing many companies, willing or not, to figure out how to pivot in a pandemic. "But by God, we were not going to let it take us down or diminish who we are," recalls Fraley. "We immediately started to put plans in place, asking ourselves, 'What if the government does this? What if the industry does that? What if we encountered this?'"

Not all companies leaned into the unknown, and as a result, many companies did not survive the pandemic's grip on the economy. That would not be the case for Price, which had built its business on execution, leaning into, and even embracing, the toughest tasks first. The pandemic would become one of its toughest tests to date.

While the COVID-19 pandemic defined the resilience and grit of countless companies, including Price, few were given the opportunity to become stronger because of the pandemic's challenges. For example, with customers "locked down" in their homes and with discretionary income normally spent on activities shelved by the pandemic, large-scale online retailers drew in record sales well into the tens of billions of dollars. And as the pandemic advanced, cleaner air indoors soon became an emerging issue and an opportunity for Price.

As a self-proclaimed "data guy," Fraley was following data from leading health authorities that showed an ominous trend: the pandemic's spread around the world would soon land in North America. In the early days of March 2020, Fraley remembers, "I pulled the team together on a Tuesday, and I said,

'I think there's a good chance our ability to work from our facility is going to be compromised in the near future. Let's make sure that everyone has the ability to work from home, whether it's a laptop, a home computer, VPN, and whatever it is; let's talk about how we would try to keep the plants operating.'" Within days after that team meeting, Fraley recalls the governor declared a state of emergency and a shelter-in-place order in the state of Georgia.

Soon afterward, Fraley and Gerry campaigned Georgia's governor to keep Price's plants open in Atlanta. "As an essential business producing product for the critical environments segment [such as hospitals] we were given those exceptions," says Fraley. Though it was a considerable amount of work to establish enhanced cleaning and sanitation processes to encourage employees to come to work, Fraley remarks proudly that they did. "We have such a devoted and dedicated workforce."

With facilities open and staff in place, the commitment to execution began in an unprecedented way. "We put together a group of five sales reps who were active in the critical-care sector and that understood the importance of HVAC's ability to distribute cleaner air. That group helped define the market opportunity as we developed our first room air purification unit. Furthermore, this group influenced how the unit would function and the features it would have," assures Fraley.

Within three months, Price had its air purification product in the market, a turnaround time that was ambitious based on the very standard of industry lead times the company helped to create. "The initial product-development initiative, and the genesis of the Room Air Purifier (RAP), was heavily influenced and driven by Nolan Hosking, Senior Vice President and General Manager, Air Moving Division, and his team," credits Fraley. "We gathered reps together and facilitated communication, but Nolan and the Winder, Georgia, team developed, tested, iterated, and presented the RAP to market. The RAP drove important revenue early in the pandemic. Other product designs would follow, but none have had the impact of the RAP. And to this day, our reps acknowledge and appreciate that we were able to step up," claims Fraley. "That product generated about $25 million in sales over a short period."

Two notable examples of products produced during the pandemic pivot included the Puraflo Fan Filter Unit (FFU) and the Ceiling Air Purifier (CAP). Both units suck in room air and filter it through a high-efficiency particulate air (HEPA) filter to remove 99.99% of particulates and then return the clean air to the same room. At that high degree of effectiveness, it was a desirable product for hospitals—pandemic or no pandemic. And the FFU removes harmful particulates that can get drawn in from a building's airflow. "Our response was to figure out how to stay in the game," recalls David Surminski, General Manager of Price's Critical Environments business unit. "We wanted to serve our customers, keep our people and our communities healthy, and our countries well positioned to exit the pandemic. We weren't narrowly focused on beating our competition or making sure our product was 'better' than someone else's. We took the long view and figured it out."

Hitting a home run

Price's long-established culture of execution allowed the company to not only survive but thrive during the pandemic when it came to product development, by launching intricately engineered air purification products within months of conceptualization and producing stock with intensely demanding timelines. "Gerry created an environment of expectation. It's just his nature to say, 'We've got to get this done. We've got to.' But the way he did it was with regular meetings, some of them called Home Run meetings, where teams reported on progress and held each other accountable in an encouraging way, and that just kept things moving forward with a targeted deadline," assesses Fraley. Home Run meetings included all the people who could contribute to the success of a project and, if executed, would have a material impact on the success of the business—like a home run's impact on a baseball game's score.

"Every single major software development has been a Home Run meeting. Every expansion to the operation has been a Home Run meeting," insists Gerry, confirming how important the gatherings have become. "It was game-changing for the company. We had many Home Run meetings running concurrently in the company." While the frequency of meetings fluctuates depending on the sense of urgency, the Home Run's objective remains the same no matter who is involved or how many, explains Gerry. "It's a no-ego environment. There's no one-upmanship. The goal is to get it right, to get it done more quickly, and to have the company succeed as a whole."

"You're not going to get there if you're not executing, and that was where we had to bring focus. We had to get certain things done because there was a set of sequential steps that we had to work through to get a finished product to market. Getting the team aligned and getting them to attack the project with urgency while remaining diligent was critical. And that's just what they did," recalls Fraley. Reflecting on the pandemic's early stages, he adds, "Our team sat down with scientists, and collectively, they quickly developed a prototype." Execution, claims Fraley, was what made the difference at each phase of product development: "You couldn't go to the next step without execution."

Executing an adrenaline rush

Over the decades, Price has built its reputation and influenced the air distribution segment of the HVAC industry by developing short lead times that have become the standard, most notably during challenging circumstances, including a pandemic. "I think it's an adrenaline rush for all of us," observes Cyr. "In unprecedented times, it's intense. It matters and it's personal. We want to do it. And there's just an incredible belief that we're going to do this. By the time it got to the point where we were moving forward with our COVID response, we knew we could do it, we knew we had the team, we knew we had the resources, we knew we had the labs, and that we would pull it off."

The Tenets that make up the Price Way intuitively overlap, providing a consistent source of direction, and in the case of the pandemic, *Execution* connected the *Greater Good* to *Service* and to *Innovation*. "It's incredibly inspiring and motivating to participate in something special, where you're making a big contribution, not just to the business, but to society," says Cyr.

Action is the default

Understanding the commitment to execution exists at every level at Price, most notably in the planning cycles, and from there, broken down into actionable steps and discussed in detail in meetings such as Home Runs. Execution is what powers Price through many large-scale projects. Of lessons learned from Gerry, Maykut emulates, "It's just going to next year's operational plan and identifying the top three things, then the next three steps the following week. It's just that default to action. You don't know where it's going to take you in a year or in four years, but if you just keep grinding along, doing the right things, defaulting to action in the short term—it's the right way to move a company forward."

Good plan, great execution

"Fail to plan, and plan to fail." Two major US plant relocation projects Fraley and his team managed illustrate this maxim; two contrasting outcomes show just how important planning is.

Fraley describes the first relocation outcome: "In August 2008, we moved our air terminal business out of the Suwanee, Georgia, plant into the Winder, Georgia, facility. The planning was exceptional, and the execution of relocation did not disrupt service to the customer," he remembers. "That plant started building and shipping products to customers almost immediately. The team managed to foresee possible obstacles, and things just went well for them. They had the right timing in place, the right resources in place, and everything kind of fell into sequence."

The second relocation outcome involved the September 2019 move of Price's noise control business from Suwanee into the nearby Crestridge plant. At the Crestridge facility, Price manufactures radiant and noise control products. Not all execution went as well as at the Winder location, and when it does not, it often goes back to the planning. "We did not spend enough time working with the plant group or the people that believe in that plant, and there were disconnects. It did not go as well. And it took us two years to get that plant figured out as a result," chronicles Fraley.

Empowered by trust and transparency

If Gerry were to consider a 14th Tenet for the Price Way, *Transparency* could be a serious contender. Of the stakeholders interviewed for *I'm Just Gerry,* many of them offered transparency as one of the top reasons they enjoy working for the company, and that includes Billy Eckardt. He is a Production Control Manager at the company's plant in Winder, Georgia. At 120,000 square feet, the plant moves hundreds of thousands of products every day. It is here that execution is not a nice-to-have but an absolute need-to-have. Eckardt's job is to oversee orders, ensuring they are correct and are out the door on time. "'Execution' is the key word here as we have to move the orders as swiftly as we can," explains Eckardt. "We pride ourselves on taking the least amount of time to get orders out compared to any of our competitors. We keep our lead times very short, and it's not always easy to do."

Lack of raw materials, staffing shortages, and getting trucks in the facility on time are a few reasons why execution poses a test to Eckardt and his team on a regular basis. The development and consistent review of contingency plans to mitigate challenges allow the Winder plant to remain 90% on time when fulfilling orders.

Eckardt was one of Price's first employees when the

company set up shop in Georgia in the late 1980s. "We were always kept in the dark back then," recalls Eckardt of the time before the transition to Price. "You didn't know one day to the next if we were going to have orders on the shop floor." Just a few days after Price became the new owner of the Winder plant, Eckardt knew things were going to improve for the better, and he confirms it has remained that way ever since.

Through recessions and other setbacks, both in Canada and the US, Eckardt purports Price's commitment to remaining transparent about challenges and opportunities facilitated greater communication for his teams to plan their execution. "With all the recessions we've had and all the difficulties that we've faced, we've always grown through execution."

Execution in real time

As part of its ongoing commitment to facilitating the *Execution* Tenet throughout its operations with transparency, Price has invested in key initiatives to ensure its teams remain in the know. For example, in the Winnipeg headquarters, staff from all departments will find monitors on walls in key high-traffic areas that post real-time orders received and fulfilled. "Maintaining trust and transparency and continuously communicating are paramount to the execution process," says Loeppky. "We also have placed visibility monitors throughout the factories with key performance indicators so that everyone knows how we are doing on a daily basis, for better or for worse."

●

Price Points on Execution

Pivot in a pandemic

When the COVID-19 pandemic shut down the North American economy, many companies did not survive. At Price, the pandemic would become one of its toughest tests to date. The company responded by leveraging its vertical integration with in-house parts and technology and quickly developing products to support critical environments.

Hit home runs

Home Run meetings are a way to keep projects moving forward and were a long-established practice at Price well before the pandemic. Home Run meetings include all the people who could contribute to the success of the project and are an opportunity for teams to report on progress and hold each other accountable in an encouraging way.

Default to action

The commitment to execution exists at every level, most notably in the planning cycles. Execution is what powers the organization through many large-scale projects as well as a crisis like the pandemic.

Build trust with transparency

Through recessions and other setbacks over the decades, both in Canada and the US, Price's commitment to remaining transparent about challenges and opportunities facilitated greater communication for its teams to plan and follow through with their execution.

# Financial Discipline

GERRYISM

Price is a non-bureaucratic, "lean and flat" company. Our leaders are all experts at one or more business or engineering disciplines so that they can drill down to the detail level at any time and gladly do so as and when necessary. Staffing is kept to a minimum, with competent high achievers all carrying a full load and relishing it.

FINANCIAL DISCIPLINE TENET

Business owners can either take profits out (for lifestyle or to diversify) or leave profits in (to fund growth and be recession proof). We choose at Price to keep the vast majority of our earnings in the company to finance innovation and start-up losses internally, plus to make our business financially strong so as to be able to survive the left curves from a wildly fluctuating economy.

The Singapore surprise

One morning in July of 1985, Gerry received a desperate call; the connection was full of static, but the news was crystal clear. On the line was Othmar Furer, Plant Manager at Price's factory in Singapore. "I just want to let you know that we're sinking fast. You'll be coming here one way or another, either to liquidate the business or to turn it around. But something has to be done," remembers Gerry, paraphrasing Furer's ominous warning.

Gerry was yet to be named company president, and he had only recently taken on the responsibility of looking after the Singapore start-up. The general manager in Singapore, who also had a small stake of ownership in the facility, had let operations slide substantially. As no regular checks and balances, including monthly financial reports, had been established prior to Gerry's assignment, he had to act fast.

Gerry had not been aware just how dire operations had become. "I went into overdrive," recalls Gerry, "shifting my life and that of my family into high gear. I'd already paid the fees for our girls' school activities for the next year. They were 10 and 12, and I said to Barb, 'Would you like to go to Singapore?'"

Thinking Gerry was offering an enticing family trip overseas, Barb enthusiastically accepted, embracing the experience and adventure for her young family. Within a few weeks of Furer's news, Gerry and Barb rented their Winnipeg home to Barb's brother, pulled their daughters' school registrations, and were on a plane.

Upon arrival in Singapore, the dynamic duo quickly found

accommodations. "That was such an exciting experience for us as a family," says Barb, finding humor in the experience, though at Gerry's expense. He was working exceedingly long hours in Singapore, often in a plant office that was void of windows. "I remember we came home at Christmas, and everybody asked, 'Gerry, where's your tan?' and he said, 'I don't get a tan from fluorescent lights.'"

Now acting as general manager in Singapore, Gerry quickly discovered the problems did not stop with production—the sales organization was equally problematic. It had lost two major product lines that Price represented in that territory, which accounted for half of the operation's gross profit. In the following year, Gerry developed and implemented a turnaround strategy, and by June of 1986, he had replaced the problematic general manager with a new leader who was also a minority partner. Soon afterward, the sales stabilized, and Gerry purchased a factory from a bankrupt rubber company, leading the way for the Singapore plant to become profitable in 1987.

What became abundantly clear for Gerry with the Singapore purchase, and its subsequent crisis, was the importance of timely financials. The state of the financials when he visited the company in February 1985—months before Furer's desperate call—was a foreboding precursor to the crisis that followed. "I showed up in February, and the accountant said we do not have the financials ready for the prior year. When they were finally revealed 10 days later, and the day before I was scheduled to leave back to Canada, they showed a massive loss—a loss that was equivalent to the total profits made by the entire company back in Canada. A physiological wave of heat and nausea rushed through my entire body, and I felt a deep responsibility to all those who contributed to that success back home. I will never forget that awful feeling. That physiological response burned into me the importance of knowing monthly financials in real time."

According to Gerry, good news can always wait—but bad news, especially if that news reflects in the financials, cannot come soon enough. Gerry describes it in context to a left curve, meaning an unexpected and challenging event or occurrence in life requiring an immediate response or timely course of action. "When left curves happen, and you have no chance to alter the left curve or do damage control from the left curve or create a survival plan if there's no time to react to correct—it's a sinking ship and you're done."

To mitigate any future left curves, Gerry instituted flash financial reports that showed, at a glance, the financial health for all his operations. After the Singapore surprise, flash reports evolved from three weeks' reporting after a month's end to present day, which is now down to a week. The lesson learned would serve Gerry and Price well just under a decade later, this time in the Canadian operations.

You can't live off elephants alone

The other lesson learned from the Singapore surprise for Gerry and Price was shifting Western ways of doing business to foreign strategies. What worked in North America could not always be applied in Asia. Gerry quickly discovered upon arrival in Singapore that local competitors produced product at a very low cost with estimated science, locking up virtually all the small to medium contracts. What at first seemed like a blessing—Price picking up the remaining big contracts—soon became problematic when those few-and-far-between jobs negatively impacted cash flow and stability because of the long gaps between big projects. "We're coming in doing it the Western way, with quality science and first-class product, that requires a higher price point in the market," remembers Gerry. "We could not get to the price point of the local competitors, and I didn't realize that until I got there. By then, we had already launched our current line of more expensive product. We were living off of the elephants—the big, the prestige, and the quality jobs— and almost nothing of the small to medium jobs. The lesson from Singapore was you can't support a factory living off of elephants and buffaloes; you need the rabbits and the mice and the squirrels—even the ants and little stuff, so to speak."

Lesson learned for Gerry and his Price team in Asia. "It was burned into me in the 1980s that to be successful in a country as a start-up, you must be a low-cost producer, with a product that could go head-to-head with a local low-cost producer and be viable on the small to medium jobs. If you get the big contract, that's the gravy. However, if you can't compete on the small and medium-size jobs, get out of the market."

The Canadian collapse

The Canadian collapse of the economy in the late 1980s, leading into the recession of the mid-1990s, decimated many industries, including the HVAC industry. According Price's estimation, the market size went from 180 million square feet of non-residential construction in 1987 down to 60 million square feet by 1993. That is approximately a two-thirds loss of non-residential construction opportunity in Canada. This was particularly troubling as it was the Canadian side of Price funding the start-up losses in the US expansion.

"So the question is, then what do you do? Well, because the mid-February flash financial report in 1992 told me we were in trouble in January, I waited for February's flash in early March to confirm it," recalls Gerry of the good news/bad news scenario. The good news was the flash reports eliminated a surprise left curve; the bad news was that the Canadian operations suffered two months of decline. "Thanks to relatively real-time awareness of the collapse in the Canadian market, which was the primary funder for our US losses, I went into overdrive on a survival plan for Canada," explains Gerry.

Salary wages were rolled back, including unionized factory wages, a normally unprecedented labor-relations move. Despite the desperate circumstance, where emotion could have easily trumped reason, Bill Cross, a longtime factory-floor worker at Price and who also doubled as a union negotiator, recalls how tactful and respectful Gerry negotiated with the union. "Talks were not going well. It was a tough time for the company. Gerry invited the negotiating committee to come meet him downtown, and he laid it all out there for us. He was not demanding. He was just speaking from the heart, saying this is what we need to do to keep the company afloat."

Executive salaries were rolled back, and dozens of people were laid off at the Winnipeg plant. Additionally, company pensions and benefits were suspended, with the company shifting into an extreme austerity program.

The program worked. Price lost less in the next quarter than anticipated, and earning profits were up by the next two quarters. Thanks to real-time financial reporting, Gerry was able to swerve the left curve of bankruptcy into solvency, staying viable. "My first real incentive to institute timely financials with flashes was in Singapore, and I translated that to the Canadian operations, and thankfully in time to save us in the Canadian recession," says Gerry with relief while offering a glimpse of what could have been without flash reports. "The economy can turn on you in a heartbeat, and you won't know it until you're sunk."

A debt of gratitude

An interesting revelation about the truncation of the Canadian market for Price is that the market was already showing undisputable signs of peaking in the 1970s and 1980s. According to Craig Brown, Price's longtime Chief Financial Officer (retired December 2022), the Canadian recession essentially hammered the final nail in the proverbial coffin. "Non-residential construction opportunities had shrunk considerably in Canada, and it never really recovered."

Gerry often refers to that recessionary period as the "de-industrialization of Canada," which Brown asserts is completely accurate by his assessment. "Our Canadian non-residential numbers today, even with inflation, don't match what they were back in those heyday years. There's no way we would have been able to scale up, and with the recession, we most likely would have gone under at some point."

Thankfully, Gerry had his blueberry pie to plunder—US expansion—which facilitated massive scalability. The expansion, as you may recall in the chapter on Risk, was not well received by Gerry's key stakeholders. With substantial losses mounting in the US and the Canadian recession making matters worse, Gerry took on dizzying levels of additional debt with loans to pay out those with a stake in the company who could not cash out soon enough.

In today's banking environment, where relationships between companies and banks are more pecuniary in nature, Brown doubts Price would receive the loans it did back then when the company was enduring such cataclysmic circumstances. "Gerry structured the finances so that he was able to do what he needed to do to keep this business alive and make it grow. And if he hadn't made that decision and we hadn't moved into the States, we probably wouldn't exist today."

Banking on future success

By the mid-2000s, Gerry's debt-driven destiny had become all but a distant memory; Price was now creating millions of dollars in profit (instead of debt) and reinvesting it back into innovation and start-ups. In fact, a whopping 80% of its profit was being reinvested.

"By 2014, our bank was suddenly asking some pointed and difficult questions. They advised they were not so sure about our current reinvestment strategy in innovation," reflects Cyr. He broached the topic of the bank's concerns with Gerry, who responded, "Banks impose a certain level of discipline. There's a reason why they need these debt-to-equity, debt-service, and other ratios. Recalling Gerry's directive, Cyr notes, "There's a reason why they're asking these questions, and Gerry realized we needed to steer a bit more to the middle of the road, balancing investment in innovation, with a focus on execution and operations' strength."

In essence, this was an exercise in financial discipline, and Gerry was a keen student, and it paid off. As a result of taking a positive and constructive approach with respect to the banks' obvious nervousness, Gerry continued to invest in innovation and start-ups, but more in alignment with financial best practices designed for the company's long-term benefit.

"He wasn't annoyed. They had a point. We had just been investing so heavily and so rapidly that Gerry made a very conscious, deliberate decision to steer more toward the middle of the road. Now, what we would call 'middle of the road' is maybe closer to the fast lane for others. This was about making sure we're not so out there," offers Cyr.

The company scaled back or slowed down on several major product lines to focus more on execution for a period of time. "It worked stunningly well," boasts Cyr. "Subsequently, 2015 through 2022 have been the best years of growth and financial performance in the history of the company." Banks and businesses can, and often do, have love-hate relationships, and where many owners may have resisted advice from a banker, Gerry's legacy for a forever company includes "don't risk the mother ship," and the bank was advocating for the same financial stewardship, just in different words. "It's best to pay attention, listen, adjust, and not get emotional," says Cyr.

Don't risk the mother ship

Although Price survived periods of almost inconceivable risk, Gerry's intention was never to grow Price in a wanton manner. Legacy and the viability and growth of the next 75 years are what matters most to Gerry now, and to ensure that legacy, it is important to protect the organization, to make it a forever company. To do so, it is important that leaders tell themselves, "Don't risk the mother ship."

"That includes not taking balance-sheet risks and remaining cognizant of any and all financial consequences." Gerry further qualifies that risk will always remain part of the financial equation. "I kind of thrived on risk, you might argue—or at least uncertainty that needed to be made certain."

If the safeguard of the mother ship ever appears uncertain, falling back on financial discipline was what Gerry resorted to— even if it was an unorthodox strategy based on a background of modeling and applied mathematics. Cyr notes, "I think it needs to be understood that it sounds like or would seem like the company was at risk, but it never really was. And the reason that it never really was is Gerry's understanding of the financial reality of the business. Every year, he works up a forecast and predicts the year's financial outcomes. And how close it comes is uncanny." As part of his legacy plan to protect the mother ship, Gerry cascaded his idiomatic financial understanding of Price down through to his leadership team, that team's direct reports, and so forth to other leaders, not only for the company today, but for the company in the future.

●

Price Points on Financial Discipline

Have timely financial reporting
Good news can always wait—but bad news, especially if that news reflects in the financials, cannot come soon enough. Knowing monthly financials in real time is essential. The sooner you know your last month's financials, the quicker you can respond to crisis.

Secure small to medium contracts as a start-up
A proven strategy to succeed in a foreign market as a start-up is to be a low-cost producer with a product that can go head-to-head with a local low-cost producer. The objective is to remain viable with small to medium jobs; getting the major contracts is a bonus.

Use financial reporting to calculate larger issues
Establishing timely financial flash reports saved the company's Canadian operations during the Canadian collapse of the economy in the late 1980s to mid-1990s. The flash reports showed real-time awareness of the implosion of the Canadian market, which gave the company the information needed to create a survival plan.

Don't risk the mother ship
If the safeguard of the mother ship ever appears uncertain, fall back on financial discipline. That includes not taking balance-sheet risks and remaining cognizant of any and all financial consequences.

Build for the future

If the company remains consistent with knowing when and where to take on risk, its chance of becoming that forever company will remain in sight. By not putting banking covenants at risk or taking on more debt than the organization can handle, an organization can become a forever company.

Distributed Leadership

GERRYISM

We always bet on the collective ability of the rank and file over the talent of a single superstar.

DISTRIBUTED LEADERSHIP TENET

It is the responsibility of every leader at Price to nurture others to lead in their own right. This leadership culture promotes trust and accountability and distributes authority and responsibility. An army of leaders will always outperform an army with a single general.

The ROI from learning by doing

Marked on his calendar for weeks in advance, the quarterly report update meeting would be in front of peers, leaders, and executive management—making the thought of presenting all the more daunting for Graham Fediuk, Director of Engineering Services at Price Research Center North (PCRN). In 2012, the young product designer had been with the company for only a couple of years, so he was relatively new to Price.

The tension mounted.

Feelings of dread and demoralization, instead of confidence and anticipation, hijacked Fediuk's stream of consciousness as he drafted his presentation document for that meeting. He was grateful for his immediate supervisor's support, but now Fediuk had to account for a major product-development project that missed the mark. Now over a decade later, promoted through several roles to director of engineering services at PCRN, Fediuk will never forget how knotted up inside he felt preparing for that meeting. After all, this was his first big project as a lead, and the first time he would give a project update at a quarterly meeting. "I was both scared and nervous. I think a lot of us are wired that way."

Design the next-generation fan-powered air terminal from scratch and make it breakthrough, the best in the industry—that was Fediuk's assignment as a very young engineer barely out of school. "When we're pushing the boundaries on innovation and trying to come up with new products with new technology, being the first to market pulls in a lot of risk," advises Fediuk. "We're not doing something that has a proven solution. We're not doing something where we can go online and find an answer. We're learning and we're trying new things on our own."

The end goal was to produce a unit that could be used for airflow for cooling or heating specific spaces. On most large non-residential buildings, there is a much larger unit on the roof that distributes the air throughout the building. However, for maximum efficiency, smaller units are required to provide ultimate air quality comfort for people inside, especially in condensed areas like boardrooms and cubicles.

Drawing in several team members from across the company to contribute, the product development was resource heavy, requiring extensive market research, design work, and preliminary testing. Nine months into the project, with a series of performance targets not meeting expectations, Fediuk faced the brutal facts. The unit, as is, was failing. It was noisy, which would annoy people. It was also spewing condensation, which could end up in ductwork, creating costly repairs for customers and an unhealthy work environment.

The day to present a project update to Gerry and the rest of the leadership team had arrived. As Fediuk walked into the packed boardroom, the normally casual glances upon entry suddenly felt like a piercing and glaring examination. "I had to explain how this project was off the rails and how we had to go back to the design phase and restart," shares Fediuk.

While Fediuk cannot remember Gerry's exact response, to this day, he will never forget how he felt. "There was no negativity that came from Gerry and his mind went to, 'First of all, that happens, and it's happened to everyone who's done a design project.' And secondly, 'What did we learn from it and how do we make this better?'" After Fediuk's brief account of how his team would move forward, the report meeting simply moved on, business as usual, much to his enormous relief. "I think it took me a few days to really digest how that went down," says Fediuk, hearing the tension from his voice relax once again, recalling that encouraging encounter with Gerry.

When asked how empowerment and encouragement in the face of failure creates return on investment for product development, Fediuk's response is twofold. "My teams have the same trust and comfort level to try new things as I do, thanks to Gerry's comments to me. If we were a company where we weren't allowed to take risks or we were scared to take risks because of failure, we wouldn't have come up with some of the new technologies and innovations that we have now. It's made us better as a company because we have better products, more innovative products, and that's because people can try new things and take risks on new designs and new technologies, knowing that they have the support of the company." In other words, or Gerryism, Fediuk's experience was the "Tuition of Innovation"—which means the knowledge gained from learning can often be applied to projects in the future.

As a keen student of innovation, Fediuk kept going, eventually designing and producing one of the quietest and most unobtrusive products in the air distribution market, according to Price leadership. It was quickly embraced by customers, generating millions of dollars in annual sales today.

Throughout this book, you have read many examples of the value of starting over, by learning by doing and by manifesting greater ROI through those restarts. In Gerry's words, "We don't look back and worry about money spent; we're only looking forward on how to do it better. That mindset permits a rapid pace of iteration and a rapid pace of learning by doing because we don't slow down."

Learn by doing to build the right team

Aside from hiring the right people at the right time, how does a leader take a company, Price and its affiliates, from $30 million in sales to now over a billion dollars in sales annually? "I came into business with no business training, no accounting training— nothing," says Gerry, reflecting on how he has underpinned the foundation of his company with a history of learning by doing. "I was always an applied mathematician, a very theoretical guy who was also handy. I could build anything with my hands: carpentry, plumbing, electrical, all that kind of stuff."

Gerry's first foray into Price in 1977 was trial by fire and learn by doing. Gerry was appointed President of Price Acme— the ceiling and interior-finishing arm of the mother ship. "I just operated it based on my own thinking as an engineer. I had no business conversations with my father, ever. All I know is you fail, you figure it out, you try again and find a better path. You figure that out, you tweak your path, and you try again," attests Gerry, summing up his business journey succinctly. "My whole life I learned by doing."

Just about every case study or example in this book is a derivative of learning by doing and a precursor to understanding the Price Way's Tenet of *Distributed Leadership*, empowering those who start out producing and innovating HVAC products to become industry leaders.

Here are two further examples from Gerry: "There's nothing you can't figure out. Take our high-efficiency particulate air (HEPA) filter manufacturing today. Perfect example. We didn't get going until 2016. We had David Surminski, GM of our Critical Environments business unit, work with Milholland and Associates, a HEPA-filtration consultant. Surminski starts researching what this stuff is all about. And five, six years later, and $6 million in start-up costs later, we build our first HEPA. We didn't have a clue what we were doing when we first started."

The second example refers to Price Electronics, showcased in the Innovation chapter. "Back in the late '90s, we had a smart diffuser, which needed a circuit board. The circuit boards were coming out of Tampa in lots of 100. We were throwing away more finished product than we were using because the design was evolving so quickly. That made no sense to me—total waste. I figured—well, we'll build them ourselves. I didn't know how to build a circuit board. I hired someone who did, Mike Nicholson, fresh out of school."

Into the hinterland

A foundational strategy for Gerry's US expansion was recruiting and retaining US sales representatives, and he applied the Tenet of *Distributed Leadership* with the reps just as he did with his internal support. In Gerryism terminology, it was the hinterland strategy, introduced in the Risk chapter. "We stayed clear of the big markets in the US for many, many years. In our

US start-up, we found reps who were hungry for a new line of air distribution products. They were maybe third- or fourth-stringers, or even the odd second-stringer, who was tired of their existing lines but saw a future with us," professes Gerry, recalling that once his rep team was in place, he let the reps lead because they knew the US territory more than Gerry did.

Although Price has grown beyond its hinterland strategy, when it comes to the North American HVAC industry, the approach of flying beneath the radar serves the company as it expands into new countries. "Just go to where you can win, go into where you can have success. It's okay to be under the radar.

No longer needed

With the turbulent times of the US expansion and Canadian recession behind him, Gerry's trends for growth were becoming a reality—fast. If he wanted to sustain the growth, he needed to get out of the way. "Joe, Chuck, and Marv came on in the early to mid-2000s, and they were huge contributors to the growth of the business, carrying leadership responsibility, permitting me to not be involved in everything." Gerry knew distributed leadership was more than just growth; it was the future of Price, going beyond senior leaders to up-and-coming engineers, general managers, and essential service providers. "My teams of employees were so far ahead of me. They were seeing the future better than I. Between then and now, others have stepped up, and I'm completely out of it. And boy, is that a nice feeling not having to be needed in an area where I was essential for the past 20 years."

The Tenet of *Distributed Leadership* is only as good as

its practice and, based on experience elsewhere, not easily achieved by most companies, asserts Fraley. "You must have the right environment for it. At Price, I always felt comfortable relinquishing responsibility, allowing another leader to take on more and not putting myself at risk. I never worried about that."

Price Paradox (revisited)

Introduced in the chapter on Foundation and Legacy, the Price Paradox examines the company's quest to become a "forever company" based on the sustainability of the Price Way's Tenets. Instrumental in that ambition is distributed leadership, which is why the paradox is revisited in this chapter.

"Can we transfer the secret sauce?" asks Fraley earnestly, suggesting buy-in must be indispensable no matter the generation. "I'm not diminishing that we have great people that want to go down Gerry's path and do great things with this company. But if any one of these individuals isn't embracing the Price Way, then we have the potential for some misalignment in values."

Cyr is also hyperconscious of Fraley's observation, recognizing that even though a foundation of distributed leadership has led Price through its most challenging situations and assisted in its ascension to its most profitable periods ever, the future is never guaranteed. "What better way to multiply and grow the business and give smart, well-intentioned leaders a chance?" asks Cyr. "We don't second-guess, at all. Major decisions get made, and a lot of them grow well. Occasionally they don't. We learn from them, and we carry on. It is a way to attract and keep absolutely top-notch leaders."

A voracious reader of business books, Maykut cites Simon Sinek's *Infinite Game* theory that posits that companies fail in the long term when their goals are finite—like quarterly or annual profits. Infinite goals, as defined by the Price Way's Tenets, are what will ensure Gerry's knowledge and passion will inspire generations of stakeholders for years to come.

For Gerry, the impetus of distributed leadership will continue to serve for now and forever. "I'm very confident that this train has momentum and will continue this way in perpetuity because the nature of the leaders is clear. It's the Price Way, doing it the right way. Customers are winning, and we're finding a way to market leadership. We earned the right to be the leader because we serve better than others," offers Gerry, confident in his assertion that Price's leaders and their teams have benefited from the Tenet of *Distributed Leadership* and that experience will be passed onto the leaders of tomorrow.

●

Price Points on Distributed Leadership

Learn by doing

Learning by doing is how Gerry built a billion-dollar company, inspiring and encouraging his leaders to always do the same. Within Price, this learning by doing is called the "Tuition of Innovation," and leaders know that the time and resources will pay off in the long run with better products and greater return on service.

Go into the hinterland

The hinterland strategy in the 1980s was to target smaller markets, get a foothold, and repeat, eventually moving onto other markets. The strategy worked well, as Price flew under the radar, gaining noteworthy success in smaller markets.

Commit to distributed leadership

As Price USA started to grow, the need to scale was crucial to sustain momentum. Gerry recruited experienced HVAC leaders, trained them, and then got out of their way. For this distributed leadership strategy to succeed, the right environment is required, one where leaders feel comfortable relinquishing responsibility, allowing other leaders to take on more.

Revisit the Price Paradox

The Price Paradox is that the company was built on the skills, knowledge, and leadership of Gerry Price, yet Gerry's goal for the future is a forever company. Only through the adoption of the Price Way Tenets can this transition happen. *Distributed Leadership* is among the key Tenets that will serve Price's ambition to become a forever company. As the company prepares for the years and decades ahead, leveraging Gerry's legacy, getting leaders of today and of the future aligned is critical.

Greater Good

GERRYISM

Go big or go home.

GREATER GOOD TENET

Principled businesses form the foundation of society. We do not operate for the benefit of a few privileged individuals. Rather, our goal is for everyone to benefit from our company's operation—our customers, employees, suppliers, and the communities we live in, as well as our shareholders. We give back in a material way to those in need in our communities and encourage our employees to also give back. Everyone wins!

Go big or go home

At Price Technical Center West near Phoenix in Casa Grande, Arizona, an air of excitement and buzz fills the room, energizing the many representatives and staffers chatting excitedly. All around the center, illuminated and displayed high on the walls are Gerryisms. Curious onlookers take them in, some even taking pictures on their smartphones.

Gerry stands beneath the "Go big or go home" Gerryism and captures the image; it's a personal gesture of enduring love for his son, Travis Price, and Gerry quietly shares this meaningful moment even though Travis is not there.

Travis, an active and athletic teenager, loved all kinds of sports, but one above all was his favorite—hockey. Suddenly and unexpectedly, Travis's last moments lived, in a life full of promise and passion, were while he was playing the game he loved. On Thursday, January 8, 2004, Travis, a 16-year-old powerful forward, grinding it out on the ice just like he had done hundreds of times before in competitive hockey games, collapsed in front of horrified teammates, coaches, parents, and spectators. A gentle giant at 6'2" and 192 pounds, the popular teen was taken too soon. His death was medically determined as a congenital heart defect, of which no signs or symptoms had presented themselves previously.

"It was such a shock because he was this big, strong, healthy young man. He was one of the biggest, strongest kids on his hockey team, and who would think for a second that anything like that could happen? It doesn't cross your mind. It's a blow that you feel for the rest of your life," solemnly shares Barb,

Travis's mother. "You feel it every day, but you learn how to develop a new normal. And he was such a wonderful young man. He was just loved by everybody."

Even in the darkest moment of his life, Gerry still made it his absolute priority to remain of service. "His classmates were devastated by the loss of Travis. It was catastrophic because Travis was loved and he touched so many lives. We sucked it up, first of all, to save the kids," remembers Gerry of the immediate days and weeks after the death of his son. "The first priority was no more casualties other than our son. We didn't want to see anybody else go down because of despair."

But that was not enough for Gerry and Barb. They wanted to ensure no other parent experienced losing a child in the way they had. In the early 2000s, most arenas did not have an automatic external defibrillator, more commonly known as an AED, and that included the rink where Travis died. Committing to their priority of saving lives and to Travis's fledgling legacy, Gerry and Barb led a capital campaign, with the support of individuals and charities in Manitoba, to install AEDs in as many arenas as possible.

When this piece of equipment is used by a bystander (on someone who is in cardiac arrest), the AED can restore a regular heart rhythm. In other words, it can significantly increase the chance of getting the heart beating again. This technology is commonplace in just about every public facility in North America, and thanks to Gerry and Barb's commitment to their community, their investment played a big part in raising awareness about how AEDs save lives. While they will never know if an AED would have saved Travis in that rink in 2004, they take some solace in knowing AEDs will likely spare other families the grief of losing a loved one because of the lifesaving devices.

For those who knew and loved Travis well, they fondly remembered his exuberant and uplifting personality, beaming with endless energy and empathy for people who surrounded him. Travis expressed his passion for life viscerally on almost every occasion. One saying was synonymous with Travis's perpetually positive outlook on life—"Go big or go home!" It is a saying proudly adopted and shared at every chance by Gerry, adopted as a "Gerryism," and immortalized as a promise to Travis.

The greater good gets going

As weeks turned to months, and months turned to years, the pain of losing Travis has never truly subsided, understandably for parents who suffered the "ultimate loss." Still, as Gerry and Barb grieved and their new normal set in, they leaned into their choice to seek a path of fulfillment rather than diving into the depths of despair. In the months following the death of their son, while unsurmountable at times, they pushed through their sorrow, putting on their "game faces" to elevate their commitment to the greater good, led largely through the work of the Price Foundation, and to uphold Travis's legacy. "He lived life to the fullest. I'm going to live life to the fullest, treating every day as a gift. We're here to serve," emphasizes Gerry with resolute determination.

Many successful business owners donate their resources to altruistic endeavors, and our communities are better places to live and work in as a result. For Gerry, giving back is a core characteristic of his personality. When a lack of money was a barrier to giving back, Gerry donated his time, either lending a hand with renovations or his professional expertise to volunteer on boards. "From the first day I met him, he just had this goodness about him," affirms Barb admirably.

Quietly, over the years, Gerry and Barb have donated millions of dollars to a variety of philanthropic initiatives—from building major infrastructure for charitable entities to providing hundreds of scholarships. The donations reflect their personal commitment to the greater good.

In addition to scholarships, the Price Family Foundation's two other core priorities support inner-city programs that help people access resources. These programs help move people from poverty to potential and beyond. "Once people start to realize that they count and are valued, they will have ambition, and then they start to push themselves into learning and advancing in life," insists Gerry about his duty to give back. Underserved kids and the families that support them have benefited from Gerry and Barb's generosity, done mostly under the radar and with little public fanfare.

Gerry and Barb support healthcare infrastructure, including the Travis Price Children's Heart Centre at the Health Sciences Centre in Winnipeg. "We may have top doctors, but we don't always have the top equipment. And top doctors need top equipment and facilities to deliver top services," observes Gerry, not taking his company's good fortune for granted. "I had the opportunity to advance myself and to become successful. There's a lot of people in this world that, through no fault of their own, either have a healthcare burden or were born into a family that has problems. It could be violence, it could be abuse. It could be alcohol, it could be drugs, or it could be poverty. And they have challenges that aren't of their making."

By design, Gerry and Barb have quietly supported numerous worthwhile causes over the decades, but that does not mean important stakeholders are not taking notice, especially employees vital to the success of Price. Janet Racz, Anvil's Legal Counsel and Corporate Secretary, joined Price's senior leadership after an in-depth recruitment process. A strong impression made by Gerry about his vision for the company, specifically its altruistic endeavors, has always stayed with Racz. "He's basically committed to this calling where a majority of the profits go to the Price Family Foundation and the charitable causes," remembers Racz, adding that Gerry's commitment to people over profits aligned strongly with her personal values and influenced her decision to join the business. "I've never forgotten that from the interview—his dedication to causes. It's a big statement."

Commitment to communities

To ensure Price can continue to contribute to the worthy causes, Gerry has instilled a leadership culture of long-term commitment to the communities where they operate. "I view every site we launch as a factory that should have a future. We locate a factory in a given location, we make sure it's successful, we make sure it's growing, and we find a way to enable every single factory we have to grow to the maximum." exclaims Gerry. His commitment to Price's production plant in Winnipeg is a primary example. In the mid-2000s, the Winnipeg plant's wages were significantly higher than those of its sister plant in the United States. In this financial scenario, some companies would opt to close the operations where the expenses were greater. "We placed products in the Winnipeg plant that were higher-margin products, which allowed it to stay open," says Gerry, noting the move allowed the struggling plant a chance not only to catch up but to improve and perfect the products it was manufacturing.

The sentiment of greater good extended stateside. Fraley, who leads US operations, observed the commitment to communities when times were tough, which he says is a hallmark of Gerry's character and business acumen. "He feels a commitment to the organization, to the community, and to the business environment that we're operating in; we want to be good stewards and good citizens. We're very aware of our surroundings. We put plans in place and configure them in a way that we're not doing anything that might represent risk to the environment, to the community, or those around us," states Fraley, sharing his perspective from a previous organization where it was always "profits before people."

The return on investment extends beyond the community. Suppliers and partners see confidence and value in companies that commit to operations over the long haul, creating mutually beneficial stability. The same commitment extends to employees, creating a loyal workforce. Arnhold Neufeld, a Production Cell Leader in Winnipeg, believes Price could have easily folded operations and closed the Winnipeg plant during the recession years of the 1990s. Outside of high school summer jobs in landscaping, working at Price is all Neufeld has ever known. He knew Gerry would always do what was necessary to protect the plant and its workers. "It gave me security. This is a place I'm going to make my career at. It gave me that sense of peace, like, 'Hey, he's got my back,' and as long as I want this job, it's mine," offers Neufeld with gratitude.

In the long run, it's local

The greater good starts at home. By his account, Neufeld feels his assessment that Price had the option of cutting costs and even closing the Winnipeg plant in those lean years is accurate. Eventually, the Winnipeg plant not only survived, but thrived. Purchasing cheaper supplies overseas is a business strategy Price has made almost every effort to resist. "I believe the community should benefit from our being a corporate citizen," asserts Gerry. "We try our best not to source raw materials and products from other countries, but rather support our own nation. As much as reasonably possible, we buy from North American suppliers so our kids and our families will have jobs."

As expressed in the story above and the previous chapter on Distributed Leadership, Gerry's affinity for his frontline, factory-floor staff proves that external success starts internally, with Price celebrating 75 years of service in 2024 as evidence. As ideological as it sounds, the balance between people and profits is not always easy to manage, yet it is essential to Price's model for the next 75 years. "Shareholders deserve a certain type of return, but not at the expense of somebody else," confirms Gerry. With nearly two decades of work with Price, Cyr appreciates the Price Way as the right way more and more each year. "I've lived elsewhere in that corporate world where ego and greed are rampant. People are viewed as pawns, shifted around, and lives are torn asunder. Gerry is aware of how hard people at Price work. It's central to the notion of building an enterprise that succeeds by giving back, both internally and externally."

"We must do well so we can do good"—it is a Price business objective as much as it is a motivational statement, noted by several Price leaders. And for Gerry, doing good for now and forever will always be guided by "Go big or go home."

●

Price Points on the Greater Good

Give BIG

For Gerry, giving back has and remains a core characteristic of his personality. With additional inspiration from their son who died in 2004, Gerry and Barb have donated millions of dollars to many philanthropic initiatives and funded hundreds of scholarships. Decide how you want to contribute—and then go big or go home.

Commit to the community

The leadership culture at Price is one of commitment for the long term to the communities where the company has set up for the long haul. Over the decades, economic challenges have made other factories in Winnipeg close their doors, but Gerry and his team developed strategies to keep the facility open, including placing products in the Winnipeg plant that were of higher margin, allowing it to stay viable. If you plan to launch in a given city, plan to do so long term.

Source local

Although Price does utilize overseas suppliers when no other feasible option exists in North America, the organization defaults to supporting local services and suppliers.

Gear up for the next 75 years

As the company embarks on its *next* 75 years, the Tenet of *Greater Good* will guide the company in how it continues to support its communities. Shareholders deserve a return on their investment, but not at the expense of somebody else.

 # Conclusion

"A billion-dollar company that is an overnight success story—40 years in the making," quips Gerry during speeches and presentations, usually to packed university auditoriums or in standing-room-only business events, full of people all wanting to know just how Gerry extracted his piece of the blueberry pie. However, in the quiet moments by himself or with Barb or in the company of a few trusted leaders, what "keeps Gerry up at night" at times is ensuring the Price Way remains the only way for the company.

> If I was to say the one thing I'm most proud of, it's not the wealth and success of the business; it's the way our leaders and our business operate, that our leaders operate with kindness, with high intensity, and with principle— which I think is the right way to operate in life.

> That's what I'm most proud of. It's not the billions of

dollars here and millions of dollars there. It's the lives impacted by the Price Way and that they've embraced it and they're practicing it themselves. In other words, our way of doing business can succeed. This may not be the traditional way to run a business, but it's certainly a viable way. I'm most proud of the fact that this cadre of employees we have here in all our locations operate this way, the so-called Price Way, and are successful. And others are observing that too, and maybe it's going to rub off on others. So the legacy of people operating that way is, I would say, my greatest accomplishment. By reading the book and embracing the Price Way, whether you win or lose in life's trials and tribulations, you have the potential to live well.

The overall objective for writing *I'm Just Gerry: Like an Ant on a Blueberry Pie* is to further facilitate the transfer of knowledge to future generations and to galvanize the foundation of Price in its quest to become a forever company.

Of the 13 Tenets that make up the Price Way, Gerry was asked to choose his favorite, and his response will likely not surprise you:

The number-one Tenet that will drive you to success in life is to choose to serve somebody, and in particular, your life partner and family, your employer, and customers. If you adopt that one *Service* Tenet, you will be successful in both your personal life and your business life. Whomever you're serving will appreciate it, and you'll be valued, and you'll have a good life. To serve underpins everything.

Gerry's "overnight success story" comment may seem like a tongue-in-cheek statement, but perception usually trumps reality. The general perception of Gerry's HVAC juggernaut in the air distribution sector is that Price is a US-based company with a presence in Canada. Gerry is fiercely proud of Price's contributions to the American market and, in particular, to supporting the US cities where the company's plants and research facilities exist. As captured in the many hours of one-on-one interviews with Gerry for this book, his exuberance for the growth in the US market is exceeded only by his commitment to Canada and his operations in Winnipeg. *I'm Just Gerry* is the rarest-of-rare, made-in-Manitoba business success story—a Canadian firm triumphing in business in Canada and succeeding in the ruthless US market.

Gerry confirms he *could* have shut down operations in his hometown and—given that he stood on the precipice of financial disaster on more than one occasion—he may not have been faulted for it if he had. After all, many other companies once headquartered in the River City vacated for greener pastures, even when facing only a modicum of the challenges Gerry did.

Don Leitch, a 40-year Price staffer and retired Vice President of Sales, worked with Gerry's father, Ernie, and then with Gerry. As a Price statesman who has seen it all, he sums up Gerry as possessing an overarching theme of generosity. "This probably comes partly from his father's influence, but Gerry's drive to expand into the US did not come at the cost of his commitment to Canada and to the Canadian people."

That acknowledgment and admiration of Gerry's generosity

was a consistent undercurrent flowing through the interviews completed for *I'm Just Gerry*. From the long retired to long serving and from the newly hired to recently promoted, wherever they fit inside or outside the corporate culture, the people Gerry will leave his legacy to have taken notice.

At every opportunity, Gerry praises just about everyone else but himself when the topic of the company's substantial growth and remarkable contributions to its communities arises. Yet for many of the people *he* praises, as documented in *I'm Just Gerry*, they insist if it weren't for Gerry, with his distinct vision, mindset, work ethic, and resolve to fly below the radar, to skillfully extract his piece of the blueberry pie—Price and its affiliates would not exist as they do today for countless others to experience for generations to come.

Afterword
Gerry Price

Throughout my life, I have undertaken new activities, have had setbacks that I learned from, and then have tried again, over and over, until I was successful. This I like to call "learning by doing."

As a young man, I never aspired to be a businessman. I was raised by parents who themselves lived through the Great Depression of 1929 to 1932 and World War II, profoundly influencing how they would raise their family postwar.

I was raised with the gift of independence from my parents. I had chores and responsibilities from an early age, but I was largely left on my own, to live my childhood with minimal parental oversight or involvement, as were most kids after World War II. Parents lived their adult lives and kids lived their young lives, with little overlap between the two groups. I built many things on my own, including boats, docks, and furniture, and learned new skills "by doing," building self-reliance in the process. I didn't let lack of information stand in the way of moving forward. If I didn't get it right on my first try, I'd learn from my mistakes and try again.

This carried through to my engineering education, focusing on mathematical modeling applied to numerical weather forecasting in my bachelor's and master's degrees at the University of Manitoba and PhD at Lehigh University in Bethlehem, Pennsylvania. My mathematical modeling continued into my first full-time job after my PhD at Defence Research Establishment Suffield in Alberta, modeling how the antennae on Canadian warships would respond to a blast wave from a nuclear event. I was on a scientific and research path with little oversight and considerable freedom.

In January 1977, I made a career change, leaving the security of a scientific career for the uncertainty of a business career. I was hired by Gerry Law, President of E.H. Price Ltd., to run its contracting business. At that time, E.H. Price Ltd. had been manufacturing air distribution products for the past 15 years under license to Titus, a major US HVAC manufacturer. As well, E.H. Price Ltd. had sales offices in 12 Canadian cities and $12 million in sales annually. My dad started the company that bore his name in 1949, shortly after he returned from serving in the Royal Canadian Engineers in World War II. He retired in 1972, and Gerry Law served as President of the company from 1966 to 1986. As a student, I had summer jobs working in the factory, and in the summer of 1972, I was hired to design a manufacturing software system for the Winnipeg factory, a system architecture that is still in use today.

It became clear to me early on that business was to be my true calling. I loved the challenge, excitement, and uncertainty of business. With zero business training or knowledge when I joined the company, I had to persevere through the unknowns and challenges of every new initiative. This built my most valuable life skills that have served me so well in the long term. I truly learned by doing each year in my more than four-decade business career.

On the theme of learning by doing, I would like to reflect on four key years in my business career and illustrate the many lessons learned:

- 1977, my first full-time year, serving as Vice President Contracting
- 1992, 15 years later, our highest-risk year due to massive US losses coupled with the Canadian recession
- 2007, 15 years later, the first year our US sales exceeded US$100 million, just prior to the Great Recession of 2008 to 2010
- 2022, 15 years later, the first year our personal family holding company, Anvil, had sales exceeding $1 billion, 37 years after the incorporation of Anvil in 1985

1977–Trial by fire: Learning the hard way

Not yet 30 years old, I was responsible for two fledgling Canadian ceiling and interior-finishing contracting subsidiaries of E.H. Price Ltd.—Price Acme (in the west) and Price Tri Tile (in the east). Both companies were losing money. My job was to turn them around and make them profitable. I was naive and had no idea what a challenge this would be.

In the early months, I worked hard to land contracts and track the profitability of individual contracts, increase earnings, reduce the loss, and reach breakeven. On a project we were trying to land in Saskatoon, Saskatchewan, I read in the *Western Construction News* that the job was being awarded to our competitor. I felt strongly that this was wrong and, not knowing better, called up the Department of Supply and Services in Regina to find out if the contract had been awarded. The assistant told me it was on her boss's desk for signing but was not signed yet. I asked her to ask her boss not to sign and give us a chance to install our mock-up in Saskatoon, to show why our product was superior. They agreed, so we went to Saskatoon with our products, built a mock-up, and had a session with them showing our product in comparison to our competitor's product. They agreed ours was superior and awarded us the contract. It is amazing what can be done when you have no inhibitions and do not know the protocols of whom to call and whom not to call. By reaching out one last time, we were able to turn a loss into a win.

In the ensuing months, no matter how stringently we managed the material and labor costs of sale of each contract, the resulting gross profit added up for all contracts was never sufficient to cover our corporate overheads. We were a manufacturing company with large overheads dabbling in interior-finishing contracting, a specialized business for which we had no tribal knowledge. Many of our competitors were truck-and-ladder companies with many years of experience and low overheads, often with owners supervising major contracts or on the tools themselves. By late 1977, it was clear to me that we would never be profitable. Gerry Law agreed, and my next assignment was to do an orderly liquidation—in essence eliminating the job for which I had been hired.

We could have declared bankruptcy of our two subsidiary companies, but that would have been wrong, as it would have put the financial burden of our failure on our suppliers and customers, who had done nothing wrong. The responsible thing to do was to clean up our mess at no expense to others and be as fair as possible to exiting employees. It took a full year to complete all our contracts, which involved absorbing substantial losses on several big contracts, exiting long-term leases on a negotiated basis, and transferring as many employees as possible into new positions in the company or parting with fair severance when there was no alternative.

This was my first experience with terminating an employee. I will never forget the experience of meeting with our receptionist in the Toronto office and telling her that we had no position for her. She was in tears, and I tried my best to comfort her, including offering a generous severance. I learned the hard way of our responsibility to employees to keep the business viable, so as to "not break someone's rice bowl," a phrase I learned later in our Singapore start-up in 1985. Thereafter, for long-serving employees who were no longer successful in their jobs, I would keep them on for a finite period so they could look for a job while being employed, which was easier for them than being unemployed and looking for a job.

While winding down the contracting business, I began the development of a new family of architectural products— integrated ceiling, lighting, and air distribution products that we would manufacture ourselves. I began selling these systems to contractors, developers, and owners in major Canadian and American cities. I traveled extensively, learned the pros and cons of cold-calling, developed marketing materials on the fly (as we did not have a marketing person or capability), dealt with rejection, and turned naysayers into advocates. I learned that you need to truly understand your product's strengths and weaknesses in comparison to your competitors', and I enjoyed the uncertainty and challenge of cold calls.

I didn't know what I didn't know when I joined the company in 1977; however, thanks to the multitude and variety of assignments I had in the ensuing years in all facets of the company, I was well prepared nine years later, when Gerry Law appointed me President of E.H. Price Ltd. in 1986.

I'd like to share a few of the lessons I learned through this trial by fire:

- There is immense value to diving into the most challenging and difficult problems the company is facing. You learn in a hurry!
- Parting ways with team members or cleaning up a mess can and should be done gracefully and respectfully. Remember the Golden Rule.
- Early in your career, it's all about learning, not earning. You are building a foundation for the future.

1992—On the edge of a cliff: Into the US

Many factors contributed to 1992 being the highest-risk year in Anvil's history: we entered the year at the top of our bank line of credit, losses from our US start-up in 1992 were far higher than expected, and Canada was deep in recession, resulting in a massive Q1 1992 loss from Canadian operations. It was a perfect storm—huge losses, at a scale comparable to our net worth (Anvil's book value was $3.1 million at that time)—and we had no obvious access to capital. What precipitated this crisis, and how did we survive this most challenging of times?

Our US start-up was a very high-risk undertaking. We had no brand awareness in the US, no employees, and a limited and uncompetitive product offering. We were up against 10 established competitors, five of which were of scale with the other five being legacy brands in decline or small niche players. The market leader, Titus, was part of a publicly traded company that owned five air distribution brands, with consolidated manufacturing and deep pockets. In 1989, we started out as 11th in the US market and had aspirations plus a plan to become the market leader in time.

Our weak financial state in the lead-up to 1992 was in part due to US start-up losses totaling $4.2 million in the previous three years, plus share redemptions by the company totaling $6.8 million over the previous four years (with a further $2.5 million to be paid out in the next six years). I purchased Gerry Law's shares as well as other retiring shareholders' shares. Still other shareholders had their shares redeemed by the company because they were not comfortable with the risk of new leadership, our losing the Titus license, and the beginning of an expensive US start-up. As a result, our operating line was at the top of our line of credit.

The enormous expense of preparing our company to be successful in the US, plus the losses we had from our fledgling US plant, proved to be far greater than expected. In 1987 and 1988, we went to market in the US with our Canadian-built HVAC products until such a time that we could commission a US manufacturing facility, which we knew would be essential if we were to give our customers the superior service that they deserved. We planned to sell through independent representatives (reps), who we initially met by attending industry conferences (ASHRAE shows). Over time, we made connections and appointed 25 reps in our first two years. Most were small to mid-sized rep firms, hungry and motivated to grow, with fire in their bellies, rather than the large, successful rep firms that did not need the income and had no financial incentive to do pioneer work for our Price brand. We learned early on that our high-quality Canadian products were not competitive in the US, and we would never be of scale until we systematically reengineered and retooled all our commodity products. Also, the importance of having a US manufacturing facility to properly serve our customers grew greater every day— we desperately needed a US manufacturing facility.

Putting down roots: A home for Price

Our site-selection study to determine the location of our first US factory began in 1987. We knew from our Southeast Asia start-up in the early 1980s that the best way to enter a new market is with rented premises, so as to learn the tribal knowledge needed to get the factory exactly right before designing our own facility. It took us two years to complete this study, and by late 1988, we had three candidate cities for our factory: Atlanta, Georgia; Lincoln, Nebraska; and Sioux Falls, South Dakota. Late in 1988, we learned that United Airflow, an HVAC manufacturer in Atlanta, was failing. In January 1989, we were fortunate to be able to acquire its used assets and rent manufacturing premises at a discount, including an operational paint line. Prior to United

Airflow, Trox USA had a two-year run in this same facility and also failed. We were the third HVAC manufacturer to attempt to penetrate the US market at these premises in Atlanta, and I was determined not to strike out!

The reengineering and retooling of all our commodity HVAC products proved to be a far more expensive and time-consuming process than we ever expected. We knew that to be successful and a major player in the US market, we had to be competitive in the small and mid-sized job market, as we learned from our start-up in Singapore that you can't support a factory on only large jobs. This is because there will always be a gap between large jobs with the factory underutilized, losing money at a furious pace. It is better to have a steady flow of small and medium-sized jobs that enables a factory to be efficient and viable and to occasionally overlay the odd large job as a bonus.

Our Canadian products were both high quality and expensive. However, there is no reason why a product can't be both competitive and of first-class quality. With literally hundreds of product families to be reengineered, our challenge was to decide where to draw the line in our value analysis and reengineering efforts and which products to reengineer first. Prior to 1992, we had successfully reengineered 18 major product families, and we had eight additional product-development projects scheduled for completion in 1992. We addressed the game-changing products first—those with the greatest chance of having top-line sales growth coupled with gross profit improvement. In other words, the priority was products that would best drive incremental earnings greater than incremental overhead so as to reduce our loss and get us to breakeven faster, before we ran out of money to cover our losses.

It was essential that we also grow our sales as rapidly as possible to reduce our loss run rate and get to breakeven sooner rather than later. To this end, we studied our competitors and identified their strengths and weaknesses so we could focus our efforts where they were most vulnerable. We focused on the "hinterland"—small to mid-sized markets where our competitors were less likely to notice our presence and drive pricing down to keep us out of the market. We didn't compete seriously in the large US cities until we were entrenched in the hinterland. We focused our efforts where our chances were best, below the radar of our big competitors, as the goal was to win, not to be noticed.

We also put a big effort into communicating with our customers, to learn their highest priorities and needs and to enable their growth and success. We felt that if we could help them succeed to the highest level, we as their supplier would in turn prosper. Our focus was on their prosperity, not on our own. In addition to meeting with customers at conferences and in their offices, we met with a small group of our best customers, which we called our Rep Council, once each year, during which we presented our plans for the next year and asked them to critique our plan and present their own major priorities, action items, and needs as they saw them, to enable them to prosper and grow. Each year, their top action items became our top goals to achieve. Our record of doing this was excellent right from the outset, as in our November 1991 Rep Council, the reps' number-one ask was for us to entirely redo our catalog, a three-inch-thick binder filled with brochures on our various product families, and instead make it a concise "phone book" type of catalog. We agreed and promised to complete this in one year. At our

November 1992 Rep Council, we presented our new catalog in draft form, ready to go to the printers.

Start-up tribulations: Powering through

Start-ups are expensive and start-up losses are nearly impossible to predict, both in scale and in the length of time to reach breakeven. In 1989, I predicted our total US start-up losses over two years (1989 and 1990) would be $1.3 million and we would break even in 1991. Instead, our losses were over seven years (1989 to 1995) and added up to $8 million, and we had a small profit in 1996, our eighth year in the US. In 1992, our US start-up loss alone was $1.6 million, down from the loss in 1991 of $1.9 million, but still an enormous amount given it was about one-half of Anvil's $3.1 million book value at that time. Start-ups are not for the faint of heart.

We planned to offset our 1992 US start-up loss of $1.6 million with earnings from Canadian operations. However, in Canada, non-residential construction starts were in serious decline, from a peak of 170 million square feet in 1987 to 60 million square feet in 1993, a 60-year low. As a result, our Canadian sales declined from $43 million in 1989 to $39 million in 1991. Although 1991 sales were down by 10%, market-share gains along with various belt-tightening measures kept our overhead in step with reduced sales, resulting in a reduced but acceptable bottom line for the year as it came to a close.

Early in 1992, we were surprised at the rapidity of the decline in our Canadian sales, dropping a further 26% below our 1991 pace. We learned from our Singapore start-up how important it was to get timely monthly financial reports, within 10 days of month end, so as to get an early read on changing fortunes and be able to respond decisively in a crisis. By early February 1992, we knew our January loss was $225,000 and felt we needed one more month of results to confirm this trend. By early March, our two-month loss was $460,000, which at that run rate would result in a $2.5 million loss for the year, game-changing and possibly career ending given the magnitude of our US losses. Things had gone from bad to worse.

Radical belt-tightening measures were launched. We scaled back our Winnipeg hourly workforce from the 190 workers a year earlier to 104 workers at the end of Q1 1992. As you would expect, there was great distress and concern for the future within both our hourly and salaried workforces, and this concern precipitated a strike by our hourly workers. After the strike, we reduced salaried and hourly wages, encouraged early retirements, postponed fringe benefits and company pension contributions, had layoffs and work share, and merged staff departments with operating departments to reduce the number of managerial layers. Many engineers were relocated back to the shop floor to work alongside those that they supported (rather than residing at head office), and there was no discretionary spending. The end result was that we posted a substantial $633,000 loss in Q1 1992, followed by a much-reduced loss of $71,000 in Q2 1992, and profits in Q3 1992 and Q4 1992 that were sufficient to cover losses in the first half of the year. We ended the year slightly above breakeven, thanks to a timely intervention in March 1992 and the support of our employees and customers (who delivered some much-appreciated growth in sales late in the year).

We still needed $1.6 million to cover our 1992 US losses. To this end, in July 1992, I submitted an R&D tax credit application to Revenue Canada for our R&D qualifying expenses from the calendar years 1988 to 1991. We applied in July 1992, and to my great delight, our application was approved. We received $1,035,000 late in 1992, in time to save the year. This result, plus an income tax recovery of $453,000 and Alcan-Price earnings of $339,000, was sufficient to cover our 1992 US start-up losses, so we made it through a most challenging year relatively intact and ready to begin the fifth year in our US start-up.

Leaders, take note: Lessons for start-ups

From our manufacturing start-ups in Singapore and Malaysia in the 1980s, our US start-up in the late 1980s and 1990s, and countless product launches over the past 40 years in industries that were entirely new to us, we learned that a start-up can be "predictive" or "deterministic," meaning there are well-defined steps and actions that, if done at a fast enough pace, can achieve success (breakeven) before you run out of steam (either "will" or "money"). In summary, these steps include the following:

- Learn from and profile the competition. We started many of our US product redesigns from our competitors' best products and made as many incremental improvements as we could from there. Respect and learn from your competitors.
- Listen to and learn from your customers. We served them to the best of our ability, and when they won, so did we; their greatest needs were our top goals each year.
- Know your key drivers for success. This begins with quantifying your current reality without false ego and then identifying the key metrics for success. In our US

start-up, driving top-line sales growth and incremental GP in key product lines, greater than the incremental overhead to achieve this, were two of our key drivers for success.

- Trend is everything. It is the trend of your key metrics that counts, not the absolute value of the metrics. For example, in our US start-up, a key product, PDF, had sales increasing at 30% per year from 1990 to 1996, with GP percent increasing from −15% in 1990 to +30% by 1996.

- Win where your chances are best. Our hinterland strategy was to focus on smaller, underserved markets, below the radar of our big competitors. The goal is to win, not to be noticed.

- Focus: break an enormous task into bite-sized pieces. When we entered the US in the early 1990s, we were nonexistent in every market, and the scope of the task ahead looked immense. In order to break it down to one step at a time, we chose three cities that we wanted to launch a solid Price rep in and focused on each city, one by one, until all three cities were secured. We then chose another three cities and dealt with them one by one. By breaking the enormous US market down to three cities at a time, we reduced an immense task into bite-sized pieces that we could get our head around and deal with systematically.

- Fire, correct your aim, fire again. We did not let lack of information stand in the way of our moving forward. Lacking perfect information, we made an educated guess and moved forward rather than delaying action.

- Operate in real time. Short lead times were essential to

give our US customers the extremely good service they deserved. Fulfilling urgent customer needs instantly required the plant to be "current" at worst and "ahead" at the best of times. This meant the backlog had to be kept at a minimum all the time. We operated on the premise that "backlog is bad." It's not the big that eat the small; rather, it's the fast that eat the slow! A sense of urgency and immediacy is key.

- Review financials in real time. We reviewed financials within 10 days of month end so that a crisis or sudden turn for the worse in sales, overhead, or earnings was detected as early as possible. We could then implement a recovery plan before the damage built to a level that would sink the company. Know your reality, and deal with the cold hard facts.

- Communication is key. Through our US start-up period of great and rapid change, from being a Canadian-only company to entering the massive US market, with weak financials and great risk, communication to employees and stakeholders, like the banks, was key to keeping our team and backers onside. Each year, from 1987 on, I documented our current reality in writing, without bias or embellishment (straight goods only). From this, I created a presentation, using slides in the early days and PowerPoints in later years, and took the time to perfect the message before delivering it to employees, customers, bankers, auditors, lawyers, and other outside interested parties. This process of putting in writing our current reality and plans going forward each year, in great detail, served me well as a planning tool. It also served to keep everyone in the loop on our current reality and plan for the coming year through a much-extended perilous time in our company's history. We do this to this day.

2007—When visions comes true: Service wins

We were excited to see our US sales in 2007 come in at $109 million, significantly up from $50 million in 2003, positioning us as number two in the US HVAC air distribution market, not too far behind Titus, who was number one at that time (and number one since the mid-1980s). Remarkably, we delivered this significant growth while maintaining our extreme focus on customer service—very short lead times, most products available on two-day or one-week quick ship, and on-times greater than 95%, often at 99%.

How did we more than double our sales in only four years, with no deterioration in service metrics? The answer is threefold: (1) we invested in the factory in advance of demand so as to be able to increase sales without undue strain; (2) we significantly strengthened our US rep force, which increased our market share in our legacy air distribution products; and (3) we launched countless new engineered products, which provided additional sales growth.

In the 1990s, our sales grew at 30% per year compounded, and we invested in production capacity ahead of demand as best we could to maintain our service metrics while we grew. We had a batch-and-queue old-school shop, making it difficult to support significant sales growth without impacting our customer service. This all changed when Marv DeHart, a world-class expert on lean manufacturing, joined our company in 1999. He brought to us his wisdom and experience in lean manufacturing, continuous improvement, root cause analysis, and single-piece flow lines, which when properly tooled and configured can support very high growth rates without any deterioration in service. In the years 2000 to 2003, under Marv's leadership, our shop was completely reconfigured using lean manufacturing and flow lines, a significant investment that enabled the more than doubling of our sales to over $100 million in four years.

The significant strengthening of our US rep force was led by Chuck Fraley, an accomplished and proven senior sales executive who joined our company in 1997. Chuck had an excellent reputation in our industry and good relationships with many rep firms, and under his leadership, we managed to significantly strengthen our rep force. Many of our long-standing reps grew their way to number one in their markets, and we made changes where necessary to new reps that were or became number one in due course. This strengthening of our rep force, coupled with our factories delivering quality products and excellent service, resulted in a significant increase in the market share of our legacy products, resulting in sales over $100 million in 2007.

The final factor contributing to our rapid increase in sales was incremental sales from our launch of countless new engineered products. Since our US start-up was now profitable, we could shift our focus from survival in the 1990s to growth in the 2000s. We began a massive and relentless product-development effort that, from 1997 to 2007, produced over 140 entirely new products or reengineered legacy products, including products redesigned for manufacturability.

Alf Dyck, our chief engineer and a world-class expert on

air distribution, and I went to Germany in 2002 to attend the ISH conference, a major event held every two years. This visit, plus tours of some European HVAC manufacturers, brought to our attention the many quality air distribution products popular there. We created our own version of these products, modified to suit the unique needs and practices in the US. As well, we expanded Price Research Center North, from three research chambers in 1999 to eight chambers in 2007, to get the science right for our new families of products. We developed high-speed R&D techniques by applying lean manufacturing principles to R&D, and we automated our new labs so that testing could be done 24 hours a day, seven days a week. By so doing, lab data was converted to catalog pages in record time.

Joe Cyr, an accomplished and senior operations executive, joined our company in 2004 and played a strategic role thereafter in leading many of our growth initiatives. I remember our decision in late 2004 to go into the noise control and silencer business. We needed a sound lab, and not knowing the cost, Joe and I gave our senior engineers a one-page authorization to design and build a sound lab for $1 million. They did so, and three years later, we had a world-class sound lab at a cost greater than $4 million, complete with high-speed R&D protocols that delivered an entire family of sound attenuators and silencers, that allowed us to go to market with a complete family of products. It cost us far more than we expected, but as always, we learned by doing and became experts in this industry that was new for us. Today we are a market leader in the noise control industry.

Completing critical tasks: Rapid completion for rapid growth

This first decade of the 2000s was a period of rapid completion of critical tasks, whether it be production challenges in the factories, new-product development, customer-facing software, or other strategic start-ups and projects. Some techniques we used to proceed at a rapid pace were the following:

- Annual operational plan. Each year, we worked to a detailed annual plan complete with concrete action items, champions, and quarterly reviews. At year-end, we would do a post mortem to document accomplishments and lessons learned. We would repeat the process the next year and prepare an even better operational plan. We were not fussed about three- and five-year plans because circumstances change so fast and we didn't find they were of value. We still feel this way.
- Thrive on detail. All our leaders thoroughly understood every detail that could influence a successful outcome. It's said that "the devil is in the details." I think that's where the gold is!
- Build an ever-stronger team. When we could afford it, and as opportunities to upgrade our leadership came our way, we built an ever-stronger team. We continue to do so.
- Home Run meetings. We found that game-changing objectives can be best achieved in a team environment that we call Home Run meetings. Key team members, including senior management, met weekly, monthly, or quarterly to track performance and deal with roadblocks in real time to ensure completion on schedule. Everyone had an equal say in these meetings; there was no hierarchy. At our peak, I was but one of many on 15 concurrent Home Run meetings, some meeting weekly but most

meeting on a monthly or quarterly schedule.

- Do the most difficult first. I strive to do the most strategic, difficult, or distasteful tasks first, leaving the more enjoyable tasks for later, as time allows (many of which never get addressed, because they probably weren't that important).

2022—Following the Price Way: Reaching $1 billion

To everyone's surprise, including mine, our sales in 2022 increased 40% in one year and passed through $1 billion. This was not a goal, as our focus each year was solely on building foundation for growth to maintain our history of doubling our sales every seven to eight years. Suddenly, we found ourselves passing through a huge milestone in sales, so it's worth reflecting on how this came to pass.

The significant growth in our HVAC business over the past 15 years since 2008 was fueled by our legacy (pre-2000) products gaining another 10% or more in market share, plus our HVAC start-ups from the early 2000s being now well past their incubation stage and in their high-growth stage. In addition, we continued our relentless product-development effort from the early 2000s and produced over 520 entirely new or reengineered products in the 15 years from 2008 to 2022. In support of this massive product-development effort, we added eight new research chambers to Price Research Center North, increasing from eight chambers and 12,000 square feet in 2007 to 16 chambers and 25,000 square feet in 2015.

APEL, our aluminum-extrusion business, also grew massively in the last 15 years. APEL expanded from a one-press extrusion operation in the 1990s to a two-press operation in 2001. The decade of 2001 to 2010 was the period during which APEL solidified its foundation, upgrading both extrusion presses as well as its paint and anodizing lines. By 2010, APEL was ready to grow. APEL commissioned a new manufacturing facility with a third press in Oregon in 2010, followed by a fourth press in 2014 and a fifth press in 2018. As a five-press operation in 2022 complete with state-of-the-art finishing and anodizing, sales increased sevenfold in the 15 years since 2007. Three years ago, APEL began a major expansion, building a 330,000-square-foot factory in Phoenix, Arizona, to house three additional extrusion presses plus anodizing and a vertical paint line. This added capacity is expected to double APEL's business by 2028.

Our window business, AROW Global, also grew massively in the past 15 years by significantly increasing its market share in the bus-window market, plus by adding driver-protection products and new window systems for off-highway vehicles and service vehicles. As a result, its sales increased fourfold in the 15 years since 2007. To better serve its customers, AROW vertically integrated in glass fabrication, adding tempering to its in-house manufacturing capabilities, to give customers the shortest possible lead time.

Investing in service has driven our growth—this is what we call building foundation. We invest in people, new products, laboratories, factories, manufacturing, automation, software, marketing, and teaching aids that help us to serve our customers to a level far better than that of any of our competitors. We do not invest for our own self-interest or to increase our annual earnings; rather, we invest in the customers' self-interest, which we call "return on service," or ROS. If we get the foundation right in delivering good customer service, our earnings on a long-term basis tend to take care of themselves, even if they are occasionally low or nonexistent as we grind through expensive start-ups.

We are builders by nature. "Greenfield" is our preferred method of growth. Labs are built to get the science right, new products are designed, manufacturing facilities are commissioned, and production lines are tuned to increase capacity and margin to be sustainable. Service fuels our growth and takes us to market dominance in time. We develop new products in any industry or area we think we are capable of mastering. New products are nurtured through the high-loss incubation stage until they go through an inflection point and enter the most enjoyable high-growth stage, at some point becoming self-funding. Our decision to proceed is based not on whether the economics of the industry are healthy or not, but only on how the investment will help us to better serve our customers. We keep the bulk of our earnings in the company, to cover start-up losses, and do not invest any faster than our ability to secure funds from banks at best-customer rates.

Over the past 20 years, we have vertically integrated our manufacturing to build for ourselves products that might be otherwise available from a third-party supplier. By vertically integrating our manufacturing, we can guarantee the quality of our end product and provide very short lead times. We have become a virtual one-stop shop for the products we supply and, in fact, are our own supply chain. Examples include electronic circuit boards, hydronic coils, aluminum extrusions, metal fabrication, finishing, anodizing, specialized machines, tools and dies, glass fabrication and tempering, software, and marketing, to name a few.

We are taught that failure is the best teacher, but you learn a lot through success as well. During this period of remarkable growth and success, we learned several lessons:

- Growth is fundamental to success, and to grow you need to keep investing ahead of the curve and keep ratcheting up the scale and pace of investment.
- Growth creates opportunity and is the best way to attract and keep top people.
- Doing new things and forging new ground is inherently risky—but necessary. Be prepared for things to take longer and take more investment than you thought at the outset.
- Giving back is the morally right thing to do. I've always known this, and Barb and I have always followed this motto. Fortunately, the more success you have, the more you grow and the more you can give back.

Final thoughts: Legacy is all that matters

In 1987, shortly after I was appointed President, I developed eight corporate goals, complete with cartoons illustrating the intent of each goal, and communicated them in multiple sessions to all salaried and hourly employees using transparencies that year and slideshows in 1988 and later. The eight goals described how we would operate and what we hoped to achieve. The headings were market share, delivery, quality, customer focus, product innovation, maximum value at minimum cost, financial health, and participative management.

To keep these goals front of mind, we had T-shirts made for employees with key goals imprinted along with our current status. Over time, these original goals evolved into the 13 Tenets of the Price Way, a guiding document of how we conduct our business. This may not work for everyone, but it surely has worked for us. The Tenets and stories that precede this chapter emerged from the road I followed and what we learned as a team along the way.

Our business can be described as a "meritocracy with heart." Yes, we focus on extreme customer service and not on our own self-interest; however, we always act in a gentle and principled way, treating others as we would like to be treated ourselves. By so doing, we will leave a legacy we can be proud of, no matter the outcome, and we will celebrate because we did it the right way. How we conduct ourselves is more important than whether we win or lose. A life well lived is one in which those touched by your actions are better off for having known you. We choose to have our business be a positive force in the community, achieving a "greater good" outcome favorable to employees, customers, suppliers, and the community at large.

The power to fulfill our dream is within each of us. We alone have the responsibility to shape our lives. We are the ones pushing ourselves forward, and we are optimistic that the future will be better than the past. We believe in the motto "Go big or go home."

ACKNOWLEDGMENTS
GERRY PRICE

I wish to thank our past, present, and future employees, whose extreme service to customers and each other, operating in the Price Way, has created the growth and success we've witnessed to date and hope to see continue into the future. Working as a team, they've enabled us to survive many past recessions and have propelled our growth, whether in good times or bad. We stand on the shoulders of those who came before us. Thank you for your loyalty and dedication through the ups and downs, the trust you placed in the company, and for carrying more than your share of the load. You are our heavy lifters.

I would also like to thank our many long-term and more recent customers, whose success in their local markets is the only reason we've been successful. Your candid advice on what you need to succeed has formed the basis of our annual plan each year, and we take our responsibility to provide products and services to enable your success very seriously. It has been a pleasure to see so many of our long-standing customers become the market leader in their region, and we celebrate and applaud your success. Thank you for the business you've given us and for sticking with us through the many ups and downs over the years.

Finally, and most importantly, I wish to thank my wife, Barb, the love of my life, my best friend and soulmate, who has been my life partner since high school. Through all our challenges, the ups and downs and left curves that life throws us, she has been the rock and ever-loving supporter and confidant who enabled me and us to soldier on. She is the mother of our three incredible children, who themselves have given us five wonderful grandchildren, the joy from which is incalculable. I know I'm a lucky guy, and the greatest luck of all was the miracle of Barb falling in love with me at an early age and our building our life together. She truly is the wind beneath my wings.

ACKNOWLEDGMENTS
ROB WOZNY

I would like to thank Gerry and Barb Price for trusting me to tell their story.

Thank you to the Price leadership team and employees for their courage to share their stories, without which there would be no *I'm Just Gerry*.

A special thanks to Chuck Fraley for his invaluable US insight, Danica Schoonbaert for her impeccable technical expertise, Greg Loeppky for his unwavering commitment, Joe Cyr for his storytelling savvy, and Marty Maykut for his quiet confidence.

And finally, thanks to Jeanne Martinson, for her resilience and grit—the perfect companion on this business storytelling journey.

ABOUT THE AUTHOR

Rob Wozny has served in some of the most senior communications, content, and editorial roles, ranging from Vice President of Communications and Content at a major-league sports and entertainment company (True North Sports + Entertainment, Winnipeg Jets) to Senior Lead News  Anchor, Supervising Producer, and Reporter in leading Canadian newsrooms (CTV, Global).

Rob's first book, *Storytelling for Business: The Art and Science of Creating Connection in the Digital Age,* earned a finalist nomination at the British Business Book Awards in 2023.

Co-founder of Sound Strategy Communications, a Canadian communications and content firm that has served a diverse range of clients since 2006, Rob has earned the trust of prominent innovators, leaders, and business owners who regularly seek his counsel, creativity, and strategic vision.

To learn more about Rob Wozny, please visit www.robwozny.com.

TO FIND OUT MORE ABOUT
"I'm Just Gerry"

visit:
ImJustGerry.com

or contact us at:
info@imjustgerry.com